F. Raymond Salemme

W9-CFV-288

CONFERENCE PROCEEDINGS SERIES

Molecular Diversity and Combinatorial Chemistry

Libraries and Drug Discovery

Irwin M. Chaiken, EDITOR
University of Pennsylvania

Kim D. Janda, EDITOR
The Scripps Research Institute

Developed from a conference sponsored
by Cambridge Healthtech Institute

American Chemical Society, Washington, DC

Library of Congress Cataloging-in-Publication Data

Molecular diversity and combinatorial chemistry: libraries and drug
discovery / edited by Irwin M. Chaiken, Kim D. Janda.

    p.   cm.—(ACS conference proceeding series)

    "Proceedings of two conferences sponsored by Cambridge Healthtech
Institute, Coronado, California, January 28–February 2, 1996."

    Includes bibliographical references and indexes.

    ISBN 0–8412–3450–7

    1. Combinatorial chemistry—Congresses. 2. Drug design—
Congresses.

    I. Chaiken, Irwin M. II. Janda, Kim D., 1957– . III. Series:
Conference proceedings series (American Chemical Society).

RS419.M65   1996
615'.19—dc20   96–20429
                                                      CIP

This book is printed on acid-free, recycled paper.

Foreword

THE ACS SYMPOSIUM SERIES was first published in 1974 to provide a mechanism for publishing symposia quickly in book form. The purpose of this series is to publish comprehensive books developed from symposia, which are usually "snapshots in time" of the current research being done on a topic, plus some review material on the topic. For this reason, it is necessary that the papers be published as quickly as possible.

Before a symposium-based book is put under contract, the proposed table of contents is reviewed for appropriateness to the topic and for comprehensiveness of the collection. Some papers are excluded at this point, and others are added to round out the scope of the volume. In addition, a draft of each paper is peer-reviewed prior to final acceptance or rejection. This anonymous review process is supervised by the organizer(s) of the symposium, who become the editor(s) of the book. The authors then revise their papers according to the recommendations of both the reviewers and the editors, prepare camera-ready copy, and submit the final papers to the editors, who check that all necessary revisions have been made.

As a rule, only original research papers and original review papers are included in the volumes. Verbatim reproductions of previously published papers are not accepted.

ACS BOOKS DEPARTMENT

Contents

APPLICATIONS

Preface

COMBINATORIAL CHEMISTRY IS AN EMERGING and fast-moving field. It is driven by the profound expansion in our awareness of the molecular components of biology, the thirst to identify antagonists and mimetics of these biomolecules for use in biotechnology, and the realization that rational design of these new molecules is not a perfect art. As exemplified in the field of drug discovery, the traditional molecular discovery process can be slow and unpredictable. To solve this problem, combinatorial chemistry offers the opportunity to make diverse libraries of chemicals and to screen these for novel function. Diversity libraries offer the possibility of matching the diversity of biological targets with the diversity of potential molecular leads. Libraries do not guarantee more active therapeutics and other biotechnology tools, but they do increase the odds. These increased odds generate the excitement surrounding the discovery of biomolecular diversity and this new field aimed at using it, namely combinatorial chemistry.

This book brings together the work of participants at two conferences held at the Hotel Del Coronado, in Coronado, California, from January 28 to February 2, 1996. These meetings were organized by Cambridge Healthtech Institute to focus on successes, problems, and goals in this emerging field. Undoubtedly much more research is yet to come, but the speed of development in this field inspired the conference participants to assemble snapshots of their work. These papers are offered in this book as a record of current work, a stimulation to students of all ages, and a basis to build the field and increase the impact of its technology.

IRWIN M. CHAIKEN
Rheumatology Division
University of Pennsylvania
  School of Medicine
913 Stellar Chance Labs
422 Curie Boulevard
Philadelphia, PA 19104

KIM D. JANDA
Departments of Molecular Biology
  and Chemistry
The Scripps Research Institute
10666 North Torrey Pines Road
La Jolla, CA 92037

April 23, 1996

STRATEGIES

Chapter 1

Secondary Structure Templated Libraries: Mimicking Nature

Maher Qabar[1], Jan Urban[1], Charles Sia[2], Michel Klein[2], and Michael Kahn[1,3,4]

[1]Molecumetics Ltd., 2023 120th Avenue, Northeast, Suite 400, Bellevue, WA 98005
[2]Connaught Laboratories, 1755 Steeles Avenue, Willowdale, Ontario M2R 3T4, Canada
[3]Department of Pathobiology, University of Washington, Seattle, WA 98195

Nature has used a "library approach" to constructing ligands for specific receptors and enzymes by combining a limited functional diversity of 20 amino acid side-chains with a small array of secondary structure motifs–reverse turns, α-helices and β-strands. The dissection of multidomain proteins into small synthetic conformationally restricted components is an important step in the design of low molecular weight nonpeptides that mimic the activity of the native protein. Mimetics of critical functional domains might possess beneficial properties in comparison to the intact proteinaceous species with regard to specificity and therapeutic potential. Combinatorial secondary structure templated libraries provide a powerful engine for the development of novel vaccines and pharmaceuticals.

Peptides are characteristically highly flexible molecules whose structure is influenced by their environment (*1*). Their random conformation in solution complicates their use in determining the receptor bound or bioactive structure (*2,3*). Conformational constraints can significantly aid this determination (*4*). The advent of molecular biology (in particular, cDNA cloning and monoclonal antibodies) has provided enormous opportunities for structural as well as functional analysis of a wide array of proteinaceous species. The critical roles that proteins play at all levels of biological regulation provide virtually limitless potential for therapeutic intervention with recombinant proteins. However, with some notable exceptions (EPO, tPA, etc.), the therapeutic applications of proteinaceous species have been severely restricted. Proteins are subject to poor bioavailability, rapid proteolytic degradation and clearance, and are potentially antigenic. One approach to overcome these liabilities is to develop small molecule mimics. The power of combinatorial chemistry guided by our understanding of structural biology and molecular interaction is greatly accelerating this process.

[4]Corresponding author

1054–7487/96/0002$15.00/0

Results and Discussion

The Secondary Structure Approach. Our approach to the design of peptidomimetics has been guided by the simple elegance which nature has utilized in the molecular architecture of proteinaceous species (5). Three basic building blocks (α-helices, β-strands and reverse turns) are utilized for the construction of all proteins. We have developed peptidomimetic prosthetic units to replace these three architectural motifs. This is affording us the opportunity to dissect and investigate complex structure-function relationships in proteins through the use of small synthetic, conformationally restricted components. This is a critical step toward the rational design of low molecular weight nonpeptide pharmaceutical agents, devoid of the shortcomings of conventional peptides.

The Design and Synthesis of Secondary Structure Templated Libraries. Bruce Merrifield won the Nobel Prize for chemistry in 1985, for his development of solid phase peptide synthesis (6). It became apparent that a marriage of the simple elegance of solid phase synthesis with the ability to construct conformationally well defined secondary structural templates could become a powerful tool in pharmaceutical discovery. The synthesis of our reverse turn mimetic system illustrates this point. The synthetic protocols for the reverse turn template (and also the strand and helix templates) were developed to be performed on automated solid phase synthesizers, to take advantage of combinatorial techniques.

The synthesis involves the coupling of the first modular component piece $\underline{1}$ to the amino terminus of a growing peptide chain $\underline{2}$ (Scheme 1).

Scheme 1

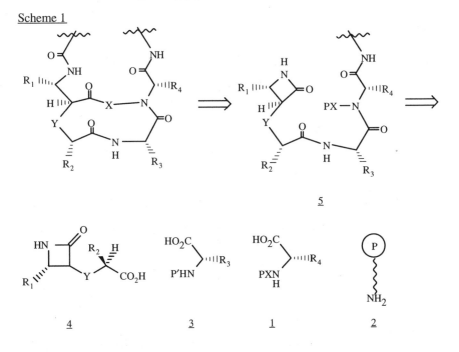

$Y = O$, or CH_2

Coupling of the second component 3, removal of the protecting group P′ and subsequent coupling of the third modular component 4 provides the nascent β–turn 5. The critical step in this sequence involves the use of an azetidinone as an activated acylating species to effect the macrocyclization reaction (7). Upon nucleophilic opening of the azetidinone by the X moiety, a new amino terminus is generated for continuation of the synthesis. An important feature of this scheme is the ability to alter the X-group linker, both in regard to length and degree of rigidity/flexibility. The requisite stereogenic centers are readily derived, principally from the "chiral pool." The synthesis allows for the introduction of natural or unnatural amino acid side chain functionality in either L or D configuration. Additionally, deletion of the second modular component 3 provides access to γ -turn mimetics (8).

Vaccine and Pharmaceutical Design

Pharmaceuticals

The Immunoglobulin Superfamily. The immunoglobulin gene superfamily comprises not only antibodies, but also a large family of cell surface molecules involved in adhesion and cell-cell recognition (including CD2, CD4, CD8, MHC, T-cell receptor, ICAMs, etc.).

Immunoglobulins are constructed from a series of antiparallel β-pleated sheets connected by loops (9,10). The specificity of these molecules is determined by the sequence and size of the canonical hypervariable complementarity determining regions (CDRs) (11,12).

Figure 1

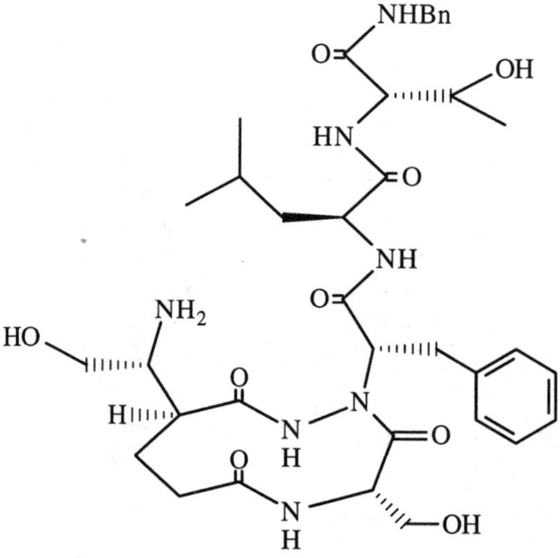

6

We have designed a mimetic of the CDR2-like region of human CD4. CD4 is a 55 kD glycoprotein, primarily found on the cell surface of the helper class of T cells. It binds the Human Immunodeficiency Virus glycoprotein (HIV gp120) with high affinity ($K_d \approx$ 1-4 nM) and is an important route of cellular entry for the virus.

Extensive mutagenesis (*13,14*) and peptide mapping (*15*) experiments have shown that the region of amino acids 40-55 within the CDR2-like domain of CD4 is critical for gp120 binding. X-ray crystallographic analysis showed that residues Gln[40] through Phe[43] reside on a highly surface exposed type II' β-turn connecting the C' and C" β-strands. Structure 6 (Figure 1) was designed as a first generation mimetic and was synthesized as previously described (*16*). NMR and molecular modeling analysis (*17*) confirmed that 6 closely mimicked the conformation of this loop. Importantly, this small molecule mimic (MW 810 as its trifluoroacetate salt) abrogates the binding of HIV (IIIB) gp120 to CD4$^+$ cells at low micromolar levels and reduces syncytium formation 50% at 250 μg/ml (*16*). An N-acyl derivative of 6 displayed 43% oral availability and a circulating half-life of six hours in a murine model.

Vaccines

Background. Peptide vaccines offer several advantages over more traditional (whole killed, live attenuated) vaccines, including simplified preparation, increased safety and decreased liability, and the ability to focus the immune response to a selected antigenic determinant of the pathogen. In practice, a major drawback of peptide vaccines is their general inability to elicit a high titer, high specificity antibody response to the native protein from which the peptide sequence was chosen. This is due to the fact that the three-dimensional conformation of the antigenic determinant is critical for specific recognition by antibodies. Due to their structural flexibility, short peptides are intrinsically not suitable for the preservation of the conformational and topological features required to induce specific high titer, high affinity antibodies.

B-Cell Epitope Optimization Using Constrained Libraries. To determine the optimum construction for the B-cell epitope portion of a tandem T-cell—B-cell immunogen, we screened a SMART™ library against a panel of V3 directed monoclonal antibodies including mAb 50.1 (*19*). Based upon our previous analysis (*20*), we have designed and synthesized a constrained B-cell epitope library incorporating both the B (*before* binding) and A (*after* binding) type constructions (Figure 2). The synthesis of the library was performed as previously outlined (*21*). The library was screened by ELISA for the ability to inhibit the binding of gp120 to mAb 50.1.
From this assay, two constructions (12AS & 14AS) were selected as candidates for incorporation into full length tandem T-B immunogens (Figure 3).

In particular, we synthesized analogs of the linear peptide immunogen CLTB 108 (Connaught Laboratories) (GPKEPFRDYVDRFYKNKRK-RIHIGPGRAF) which incorporated either the 12-membered (12AS) or 14-membered (14AS) ring reverse turn mimetic templates at the crown of the V3 loop and elicited antisera in guinea pigs against these constructions. Immunological evaluation confirmed our hypothesis that a conformationally constrained B-cell epitope has the ability to generate a high titer, high specificity antibody response. Table I shows the peptide specific antititer generated in guinea pigs after two boosts in either IFA or alum. The data show that the tandem constructions incorporating the constrained B-cell epitopes are capable of eliciting a high titer response. Particularly noteworthy is the response of the 12AS construction in alum, which at present is the only adjuvant generally approved for humans.

Figure 2

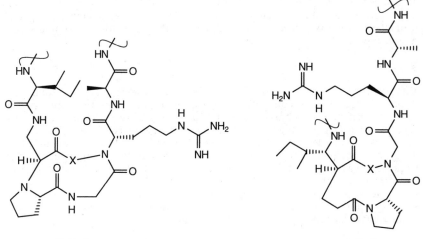

B-Series   A-Series

(Reproduced from *Bioorganic and Medicinal Chemistry*, Elsevier Sciences, not yet published.)

Figure 3

Immunogen Constructs

Constructs	T_h-cell epitope	-	*linker* -	B-cell epitope

CLTB 108   (H)-GPKEPFRDYVDRFYK-NKRK-RIHIGPGRAF-(OH)

12AS   (H)-GPKEPFRDYVDRFYK-NKRK-RIHIGPGRAF-(OH)
                                              └─12─┘

14AS   -RIHIGPGRAF-(OH)
                                              └─14─┘

**Table I. Immunogenicity of Constrained Epitopes
vs . Linear Tandem Epitopes in Guinea Pigs**

Immunization	Adjuvant	Peptide-Specific Titer		
		14AS	12AS	CLTB-108
14AS	ALUM	2,700		900
	IFA	48,600		2,700
12AS	ALUM		24,300	2,700
	IFA		48,600	2,700
CTLB-108	ALUM	300	300	900
	IFA	2,700	2,700	8,100

Gratifyingly, the response is very specific for the native protein epitope in gp120 (Table II). The 12AS tandem construction elicited exceptionally high reactivity against gp160, in both IFA and alum. In alum, the response was fifty-fold higher than the response of the linear peptide CLTB-108.

**Table II. gp160 (gp120 MN/gp41 LAI) Reactivity of Guinea Pig Antisera
Raised against Peptidomimetics and Linear Tandem Epitopes**

Antisera against	Host	gp160–reactive Titer
12AS/IFA	G. Pig	48,600
12AS/ALUM	G. Pig	48,600
14AS/IFA	G. Pig	16,200
14AS/ALUM	G. Pig	900
CLTB-108/IFA	G. Pig	12,150
CLTB-108/ALUM	G. Pig	900

The concept of determining an optimized B-cell epitope using a conformationally constrained library has broad applicability. This approach in principle can be utilized to find epitopes for either continuous or discontinuous determinants and should allow one to optimize the B-cell epitope component of a peptide vaccine in an antigen independent manner using only patient antisera (22).

High Diversity Templated Libraries. Recent advances in structural biology have highlighted the recurrent use of secondary structure motifs in interactions within structural superfamilies. For example, many peptide ligands which interact with G-protein coupled receptors adopt a biologically active reverse-turn conformation. Cytokines bind their receptors through α-helical regions. Transcription factors generally read the nucleic acid sequences of their binding sites through α-helices. Proteases and kinases recognize their substrates in extended strand structure.

Automated, solid phase modular component synthesis readily allows for the construction of high diversity combinatorial libraries of 10^4-10^6 complexities. These libraries can be constructed and screened either as spatially resolved discrete entities or as pools which can be rapidly deconvoluted (23). The novel conformationally constrained template approach allows one to use all biological data generated in the development of a 3-D structure-activity relationship (SAR).

We recently constructed a library of 4,000 conformationally constrained reverse-turn templates. This library was screened against a battery of ten G-protein-coupled receptors. From this library, lead compounds ($IC_{50} < 1$ μM in radioligand binding assays) were discovered for five different receptors. Similiar approaches are being investigated with β-strand and α-helical libraries for targets which recognize these motifs (i.e., proteases and kinases (strand) and cytokine receptors (helix)). Use of a strand library has led to the identification and subsequent optimization of an active site, nonpeptidic, reversible, potent (70 pM) inhibitor of thrombin that is in preclinical evaluation.

Although only the tip of the iceberg has been viewed, the power of combining combinatorial chemistry guided by our knowledge of structural biology is evident. Further investigations in this regard are in progress and will be reported in due course.

Acknowledgments. We wish to thank Kathy Hjelmeland for the preparation of this manuscript. MK also wishes to thank Dr. Mark Pearson for many stimulating discussions concerning combinatorial strategies and for encouragement. MK is an Established Investigator of the American Heart Association.

References
1. Marshall, G. R.;Gorin, F. A.; Moore, M. L. *Ann. Red. Med. Chem.* **1978**, *13*, p. 227.
2. Fauchère, J. L. In *QSAR in Drug Design and Toxicology*; Editor, Hadzi and Jerman-Blazic: Amsterdam, 1987, p. 221.
3. Hruby, V. J. *Trends Pharmacol. Sci.* **1987**, *8*, p. 336.
4. Hruby, V. J.; Al-Oeidi, F.; Kazmierski, W. *Biochem J.* **1990**, *268*, p. 249.
5. Kaiser, E. T.; Kezdy, F. J. *Science* **1984**, *223*, p. 249.
6. Merrifield, R. B. *Angew. Chem. Int. Ed. Engl.* **1985**, *24*, p. 799.
7. Wasserman, H. H. *Aldrichimica Acta*, **1987**, *20*, p. 63.
8. Sato, M.; Lee, J. Y. H.; Nakanishi, H.; Johnson, M. E.; Chrusciel, R. A.; Kahn, M. *Bioch. Biophys. Res. Commun.* **1992**, *187*, p. 999.
9. Kabat, E. A. *Adv. Protein Chem.* **1978**, *32*, p. 1.
10. Amzel, L. M.; Poljak, R. J. *Annu. Rev. Biochem.* **1979**, *48*, p. 961.
11. Martin, A. C. R.; Cheetham, J. C.; Rees, A. R. *Proc. Natl. Acad. Sci. USA* **1989**, *86*, p. 6488.
12. Chotia, C.; Lesk, A. M.; Tramontano, A.; Levitt, M.; Smith-Gill, S. J.; Air, G.; Sheriff, S.; Padlan, E. A.; Davies, D.; Tulip, W. R.; Colman, P. M.; Spinelli, S.; Alzara, P. M.; Poljak, R. J. *Nature* **1989**, *342*, p. 877
13. Ashkenazi, A.; Presta, L. G.; Marsters, S. A.; Camerato, T. R.; Rosenthal, K. A.; Fendly, B. M.; Capon, D. J. *Proc. Natl. Acad. Sci. USA* **1990**, *87*, p. 7150.
14. Landau, N.R.; Warton, M.; Littman, D.R. *Nature* **1988**, *334*, p. 159.

15. Jameson, B. A.; Rao, P. E.; Kong. L. I.; Hahn, B. H.; Shaw, G. M.; Hood, L. E.; Kent, S. B. H. *Science* **1990**, *240*, p. 1335.
16. Chen, S.; Chrusciel, R. A.; Nakanishi, H.; Raktabutr, A.; Johnson, M. E.; Sato, A.; Weiner, D.; Hoxie, J.; Saragovi, H. U.; Greene, M. I.; Kahn, M. *Proc. Natl. Acad. Sci. USA* **1992**, *89*, p. 5872.
17. The NMR studies were carried out on a Bruker AM400 spectrometer in either $CDCl_3$ or CD_3O^2D at $25°C$. Minimum energy conformers of the mimetic were found by a Monte Carlo search with the Macromodel program BATCHMIN (*18*). A modified Macromodel MM2 force field, in which the $N(sp^2)$—$N(sp^2)$ force constant was adapted from the Amber force field, was used for the calculations. There are two low-energy conformers, with the principal structural difference being reversal of the peptide bond orientation between residues 2 and 3. The energy difference is ≈ 0.3 kcal/mol (1 cal = 4.184 J) in vacuum; subsequent energy minimization in a volume-based continuum solution model indicates that there is no significant energy difference between the two conformers under simulated aqueous conditions.
18. Still, W.C.; Tempczyk, A.; Hawley, R.C.; Hendrickson, T. *J. Am. Chem. Soc.* **1990**, *112*, p. 6127.
19. Rini, J.M.; Stanfield, R.L.; Stura, E.A.; Salinas, P.A.; Profy, A.T.; Wilson, L.A. *Proc. Natl. Acad. Sci. USA* **1993**, *90*, p. 6325.
20. Johnson, M.E.; Lin, Z.; Padmanabhan, K.; Tulinsky, A.; Kahn, M. *FEBS Lett.* **1994**, *337*, p. 4.
21. Kahn, M. *Synlett* **1993**, *11*, p. 821.
22. Folgori, A.; Tafi, R.; Meola, A.; Felici, F.; Galfré, G.; Cortese, R.; Monaci, P.; Nicosia, A. *EMBO J*. **1994**, *13*, p. 2236.
23. For a recent review see: Gallop, M.A.; Barrett, R.W.; Dower, W.J.; Fodor, S.P.A.; Gordon, A.M. *J. Med. Chem*. 1994, *37*, p. 1233.

Chapter 2

Construct Diversity: A New Paradigm for Combinatorial Chemistry

John C. Roberts, Bert E. Thomas, Yeelana Shen,
Anita Melikian-Badalian, Paul J. Kowalczyk, and Peter V. Pallai

Department of Rational Drug Design, Procept Inc., 840 Memorial Drive,
Cambridge, MA 02139

In our approach to combinatorial chemistry we utilize pairs of highly modular and versatile building blocks to achieve structural dissimilarity at the construct level, i.e. *construct diversity*. The resulting library constructs, when derivatized in a combinatorial fashion, yield a diverse set of libraries. In the example shown, two trifunctional building blocks are connected to form four distinct cores, which give rise to 40 library constructs. We discuss the synthetic strategy and our computational approach to choosing fragments and assessing libraries.

Combinatorial libraries reported in the literature generally originate from either a core approach or an oligomeric approach. A core approach utilizes a central unit that is modified in an established way by large sets of fragments, whereas an oligomeric approach employs bifunctional subunits that are linked sequentially. We aim to exploit virtues of each of these two approaches through the use of our specialized building block set.

Well-known examples of core approaches include benzodiazepene (*1,2*) and tetracarboxylate based libraries (*3-5*). In general, a great deal of effort is invested so that the decoration of cores with a wide variety of fragments is permitted. This approach is chemistry-driven and the scaffold, structurally defined by the core and linkages to fragments, is itself a nonvariant entity. Thus, for an alternative spatial representation of fragments one has to choose an alternative core and develop additional synthetic chemistry. The lack of a systematic way to vary the library construct has been a limitation of core-based combinatorial chemistry.

In an oligomeric approach, e.g. peptide (*6,7*) or oligonucleotide (*8*), library constructs contain a conserved scaffold defined by the uniform linkages and identical distance relationship between the two reactive functional groups of the subunits. There are a large number of commercially available building blocks for some of these approaches and thus a very large number of oligomeric compounds may be synthesized. Variation of sequence space (sequence position, length of oligomer) is also possible which increases subunit versatility and compound numbers. It is the uniformity of the backbone which limits overall structural variation.

If library constructs could be easily and significantly varied, the limitations discussed for the core and oligomeric approaches could be overcome. This is precisely what we aim to accomplish with our *construct diversity* approach.

1054–7487/96/0010$15.00/0

Construct Diversity

General Concept. Combinatorial chemistry programs tend to utilize a number of library constructs, designed and developed independently from one another. This tendency reflects a prevailing sense that only a limited amount of a "total diversity space" may be covered with a given library construct. If total diversity space is loosely defined as a multidimensional description of all properties of all molecules, it appears logical that increased accessibility to library constructs leads to increased coverage of diversity space(9). Ideally, a *construct diversity* approach would provide access to hundreds of library constructs in a next-day time frame.

Modular, Double-T Approach. In an initial version of our approach, two trifunctional building blocks (T's), **1** and **2**, are connected to form double-T cores **3, 4, 5**, and **6** (Figure 1). In this example, four double-T core structures are available. These double-T cores may be derivatized at two or three sites to provide six "n^2

Figure 1. Library Constructs Derived From Double-T's

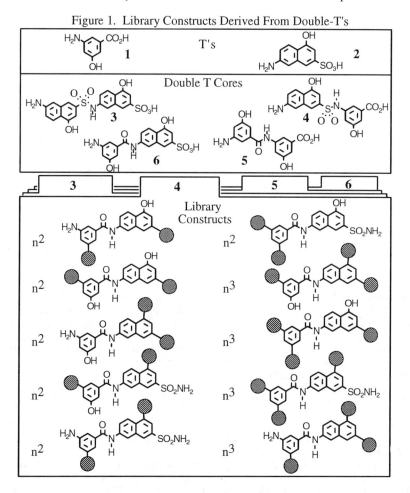

derivatives" and four "n^3 derivatives" respectively. We refer to these derivatives as library constructs, of which there are ten per core or 40 for this pair of T's.

Some of the library constructs of a given double-T core differ in significant ways. Others differ from one another more subtly (although it should be noted that both similar and dissimilar library constructs have value in lead identification and optimization). In general, one would expect greater dissimilarity between libraries derived from different double-T cores. For the sake of simplicity, let us compare single molecules. Consider the "all ethyl" derivatives **7, 8, 9** and **10** in Figure 2

Figure 2. Rudimentary Comparison of Select "All Ethyl" Derivatives

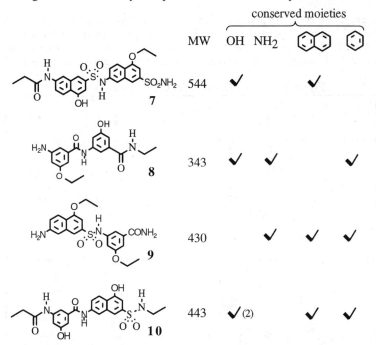

	MW	OH	NH$_2$	⬡⬡	⬡
7	544	✓		✓	
8	343	✓	✓		✓
9	430		✓	✓	✓
10	443	✓ (2)		✓	✓

which originate from double-T cores **3, 4, 5**, and **6** respectively. Decoration only with ethyl fragments creates a bias toward similarity. Simple examination of the structure, molecular weight, free functionalites, and ring systems found in these derivatives, however, indicates that they are significantly different. A computational extension of this rudimentary comparison, briefly described in the next section, would encompass a wide variety of descriptors and apply them to library members (not just the "all ethyl" derivatives).

Library Design

Fragment Selection. Once double-T cores have been identified, fragments are chosen for linkage to library constructs. Diverse fragment sets may be chosen, as outlined below.

As part of our initial synthetic approach, we link acid chlorides and carboxylic acids to amine groups, primary alcohols and primary bromides to hydroxyl groups, and primary amines to carboxylate and sulfonate groups. The Available Chemicals

Directory (ACD) served as the source for all fragments. Substructure search queries were defined for acid chlorides carboxylic acids, primary alcohols, primary bromides and primary amines. The lists generated by these queries were "pruned" based on synthetic feasibility and physical property considerations (e.g. molecular weight, solubility, etc.).

Cerius[2] 2.0 (*10*) was used to calculate a set of 51 two-dimensional descriptors for members of each fragment set. These descriptors are primarily topological in nature. To simplify the selection of diverse fragments, the multidimensional diversity space defined by the 51 two-dimensional descriptors was reduced to a three-dimensional diversity space through a Principal Components Analysis (PCA). The software allows for the selection of n diverse fragments from any data set, where n is user defined. The algorithm used chooses fragments such that (i) they are maximally diverse and (ii) the total set of fragments reflects the density of diversity space.

Diversity Measurement. We use the calculated three-dimensional descriptors to establish the diversity of chosen double-T cores and library constructs.

Here we focus our attention on ten molecules, the two T's (**1** and **2**), the four double-T cores (**3-6**), and the four "all ethyl" library constructs (**7-10**) presented in earlier sections (see Figures 1 and 2). Seven three-dimensional descriptors were calculated for each of these ten molecules: the van der Waals volume, and the average value and range of each of the three principal geometric axes. The results of these calculations are tabulated in Table 1.

Table 1. A Limited Set of Three-Dimensional Descriptors for Compounds **1-10**.

#	Vvol	AveM0	RangeM0	AveM1	RangeM1	AveM2	RangeM2
1	151.00	8.79	0.10	8.39	0.11	4.06	0.01
2	228.33	11.87	0.16	8.61	0.03	5.68	0.14
3	427.22	17.41	6.60	10.43	2.47	8.74	3.23
4	358.84	17.17	1.73	9.54	0.75	6.96	2.30
5	289.53	14.79	1.43	9.14	0.89	5.92	2.82
6	362.18	15.22	4.68	10.12	2.89	8.55	2.86
7	548.13	20.38	11.36	12.32	6.17	9.47	5.19
8	386.93	17.31	6.85	10.88	3.19	7.47	5.14
9	456.95	16.38	5.30	11.78	4.16	9.51	5.63
10	478.00	19.83	6.93	11.08	3.15	8.60	4.97

Vvol, the van der Waals volume; AveM0, AveM1, AveM2, RangeM0, RangeM1, RangM2, the averages and ranges for the three principal geometric axes.

Whereas earlier the presence/absence of chemical groups was used to demonstrate diversity at the construct level, the data in Table 1 demonstrates diversity based on calculated shape descriptors. The data were calculated for ensembles of structures and thus address the conformational flexibility of the cores and constructs.

Highlighting the data for the library constructs, one immediately sees that the shapes of the four constructs do differ from one another. Volumes differ by as much as 30%. Ranges in principal geometric axes differ by as much as 4Å for an average calculation and by as much as 6Å for a range calculation. These differences lead us to conclude that the library constructs are diverse, both in terms of present/absent chemical groups and geometry.

Synthesis

A Synthetic Approach for Construct Diversity. The linkage chemistry necessary to prepare compounds of the library constructs shown in Figure 1 is quite simple; limited to amide- and sulfonamide-bond formation, and phenolic ether formation. There are, however, selectivity issues which must be addressed. The use of four mutually compatible protecting groups would be a conceptually easy way to proceed but is practically quite difficult and lacks generality. We have chosen to develop a more general synthetic strategy, which, although conceptually straightforward, is less obvious than the use of orthogonal protecting groups.

Synthesis of a Double-T Library Construct. Combined use of (i) selective derivatization, and (ii) selective protecting group removal is an excellent replacement for complicated protecting group strategies. For example, consider the synthesis of double-T library construct **16** (Figure 3). An amino acid is attached to the resin.

Figure 3. Synthesis of a Single Double-T Library Construct

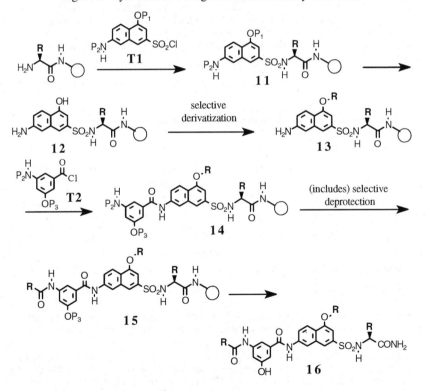

Sulfonamide formation to afford **11** is followed by removal of both the P1 and P2 protecting groups. Selective derivatization, a key step in our strategy, is then performed: alkylation of the phenolic oxygen in **12** in the presence of an amide NH, a sulfonamide NH, and an aromatic NH_2. Although we are presenting this example as hypothetical, we have achieved very high selectivity (essentially 100%) for precisely this nucleus with both alkyl and benzyl bromides. Amide formation with **T2** leads to substrate **14**. Selective protecting group removal, another key step in our strategy, albeit more routine, and acylation leads to **15**. Final deprotection and cleavage from the resin leads to the final library construct. *This combination of selective derivatization and selective deprotection allows for the synthesis of all the library constructs displayed in Figure 1.*

Conclusion

We have described our general approach to both the variation of library construct and the assessment of diversity. While we have limited our discussion to those library constructs in which one T is attached directly to another, it should be noted that bifunctional building blocks may be inserted at their juncture. This increases the power of *construct diversity* many fold. We are computationally evaluating the degree of similarity and difference between available constructs which will aid in the definition of our future directions.

Acknowledgments

The authors wish to thank Dr. Ariamala Gopalsamy for useful discussions and Martha J. Ferry for her contributions to the preparation of this manuscript.

Literature Cited

1. Bunin, B. A.; Ellman , J. A. *J. Am. Chem. Soc* **1992**, *114*, 10997-10999.
2. Bunin, B. A.; Plunket, M. J.; Ellman, J. A. *Proc. Natl. Acad. Sci. USA* **1994**, *91*, 4708-4712.
3. Carell, T; Wintner, E. A.; Sutherland, A. J.; Rebek, J. R., Jr.; Dunayevskiy, Y. M; Vouros, P. *Chemistry & Biology* **1995**, *2*, 171-183.
4. Carell, T; Wintner, E. A.; Bashir-Hashemi, A.; J.; Rebek, J. R., Jr. *Angew. Chem. Int. Ed. Engl.* **1994**, *33*, 2059-2061.
5. Carell, T; Wintner, E. A.; J.; Rebek, J. R., Jr. *Angew. Chem. Int. Ed. Engl.* **1994**, *33*, 2061-2063.
6. Gallop, M. A.; Barret, R. W.; Dower, W. J.; Fodor, S. P. A.; Gordon, E. M. *J. Med. Chem.* **1994**, *37*, 1233-1251.
7. Gordon, E. M; Barret, R. W.; Dower, W. J.; Fodor, S. P. A.; Gallop, M. A. *J..Med. Chem.* **1994**, *37*, 1385-1401.
8. Sherman, M. I.; Bertelsen, A. H.; Cook, A. F.; *Bioorg. Med. Chem. Lett.* **1993**, *3*, 469-71.
9. Ellman, J. A; *Chemtracts-Org. Chem.* **1995**, *8*, 1-4.
10. The results published were generated using the program Cerius^2™. This program was developed by BIOSYM/Molecular Simulations.

Chapter 3

Enhancing the Drug Discovery Process by Integration of High-Throughput Chemistry and Structure-Based Drug Design

T. L. Graybill[1], D. K. Agrafiotis[2], R. Bone[2], C. R. Illig[1], E. P. Jaeger[2], K. T. Locke[3], T. Lu[1], J. M. Salvino[1,4], R. M. Soll[1], J. C. Spurlino[2], N. Subasinghe[1], B. E. Tomczuk[1], and F. R. Salemme[2]

Departments of [1]Medicinal Chemistry, [2]Biophysical and Computational Chemistry, and [3]Biochemistry, 3-Dimensional Pharmaceuticals, Inc., Eagleview Corporate Center, 665 Stockton Drive, Exton, PA 19341

This chapter addresses the impact of structural data on the design and utilization of small molecule libraries as well as the impact of these high-throughput chemistry tools on the structure-based drug design process. The strengths and limitations of combinatorial chemistry and structure-based design techniques are quite complimentary. The strategies and paradigms described here in the context of our thrombin inhibitor design efforts are broadly applicable and have the potential for overcoming some of the limitations of each approach. Effective integration of these technologies will enhance the drug discovery process.

An imperative across the drug industry is to accelerate the drug discovery process. Comprehensive genome sequencing efforts, made possible by advances in cell and molecular biology, are resulting in a dramatic increase in new molecular targets that are implicated in human disease. In some cases, the most rapid and efficient way to confirm the therapeutic relevance of these targets will be through the development of prototype drugs. Many in the industry believe that the emerging technology of combinatorial and related high-throughput chemistry will impact the drug discovery process by reducing the timeline between the identification of a new molecular target and the time when prototype drugs enter the clinic (1,2). Indeed, these high-throughput chemistry methods are already finding practical application in lead generation and, more recently, in lead optimization as evidenced by the recent success of researchers at Eli Lilly & Co. who are already evaluating an orally active CNS agent in the clinic which was identified by combinatorial optimization of an existing lead (3).

The same recombinant DNA technology which has proven so useful in identifying new molecular targets can also be employed to prepare sufficient quantities of the target to support not only high-throughput screening but also structure-based methods of lead refinement. With increasing frequency, these advances in molecular biology, protein engineering, and biophysical techniques (4) (X-ray crystallography, solution NMR spectroscopy, homology modeling) are providing medicinal chemists with structural information about therapeutic targets earlier in the drug discovery process than ever before. While combinatorial strategies may prove useful even when

[4]Current address: Rhône-Poulenc Rorer, 500 Arcola Road, Collegeville, PA 19426–0107

used in relative isolation, we feel that the greatest power of these high-throughput chemistry techniques will be realized when they are integrated with other tools used in drug discovery such as structure-based design and computer modeling.

The limitations and strengths of combinatorial and structure-based discovery strategies are complementary. Although structure-based design is an intellectually pleasing approach, our inability to quantitatively predict how specific modifications of a given lead will actually affect ligand binding affinity has limited its practical application. This reflects the inability to accurately model the complex energetics of ligand binding and, in particular, the effects of polarizability, solvation and entropy. The result is that the slow, serial preparation of many "designed" compounds is still required to achieve discovery program objectives. Combinatorial and automated chemistry addresses this limitation by synthesizing readily accessible compounds in parallel so that an extensive array of structure-activity relationships can be rapidly developed. Predictions of the effects of modifying a particular substituent can then be based on empirical SAR data rather than *ab initio* or semi-empirical computations.

In contrast, limitations of combinatorial and high-throughput techniques center on the current limited knowledge base of solid phase chemistry; the dearth of reliable robotic instrumentation and high-throughput chemical analysis tools; and the rudimentary level of current chemi-informatics systems. Many of these current limitations will diminish with time as intensive ongoing development efforts begin to bear fruit. It is clear that these advances, coupled with the chemist's creativity, will give us the potential to synthesize many more compounds than it will be desirable or cost-effective to analyze, screen, track, and archive. With respect to the drug discovery process, the important question is not how many diverse examples can we make with this library, but *which* of the universe of potential library members should we prepare in order to provide the maximum useful information in the most time- and resource-efficient manner. Structure-based design addresses this problem by making the search for new leads and development of SAR more directed. Instead of producing all possible members of every library, efforts focus on the design of templates which bias libraries toward a particular target class, and the selection of specific members from these "virtual" libraries which have an increased probability of interacting favorably with the target.

In this chapter, we introduce our efforts to integrate the precision of structure-based design with the speed and sampling power of high-throughput chemistry. The purpose of this chapter is to highlight in a general sense both the impact of structural data on the design of small molecule libraries and the enhancement of the structure-based design process using these new high-throughput chemistry tools. The strategies and paradigms described here form a blueprint for overcoming some of the difficulties described above, thereby, accelerating the drug discovery process. We describe here the application of these two technologies to the challenge of identifying inhibitors for the serine protease thrombin. Active-site inhibitors of thrombin are sought after as safer anti-thrombotic and anti-coagulant agents. Despite a large potential market and much medicinal chemistry effort, selective active site inhibitors with high oral bioavailability and long duration of action remain elusive (5).

As a multifactorial serine protease, thrombin plays an important role in thrombosis and homeostasis as well as in wound healing through its regulation of the blood coagulation cascade and its ability to effect a host of chemotactic, mitogenic, and proliferative effects on inflammatory and messenchymal cells (6). Thrombin is a key mediator of the coagulation pathway by (a) cleaving fibrinogen thereby producing a fibrin clot, (b) autoamplifying its own production through a feedback loop, and (c) potently activating platelet function through its action on the thrombin receptor. Small molecule inhibitors of thrombin's active site not only provide effective control of soluble thrombin but are also capable of inactivating enzymatically active, heparin resistant, clot-bound thrombin.

Impact of Structural Data on the Design of Targeted Libraries

Knowledge of the active site architecture can aid in the design of appropriate scaffolds and templates for both combinatorial and structure-based discovery efforts. Shown in Figure 1 is the solvent accessible surface (7) of the thrombin active site. The red surface highlights the exposed portion of the catalytic triad typical of a serine protease. Natural thrombin substrates, such as fibrinogen, bind along this cleft with the N-terminal of the substrate extending out the bottom of this image and the carboxy end out the top. The subsites south and west of the catalytic triad determine to a great extent the affinity between the enzyme and its substrate or inhibitor. The major specificity pocket in the active site is the S1 binding pocket. At the base of this deep pocket is a carboxylate group from aspartic acid 189. This negatively charged carboxylate forms a salt-bridge with the positively charged basic side chain of P1 residues (i.e. arginine and lysine) thus leading to thrombin's observed substrate specificity. For this reason, this pocket is often referred to as the guanidino binding site.

D-Phe-Pro-agmatine **1** (refer to Table II for structure) served as one of our early reference compounds (8). Its determined binding mode (9) is also shown in Figure 1. In this perspective of the active site, the charged guanidine moiety of the agmatine (magenta) is concealed from view by binding deeply into the S1 binding site as anticipated. Two additional binding sites are also seen in Figure 1. The first is a large hydrophobic pocket which is shown here to cradle the phenylalanine sidechain (dark blue). This large hydrophobic pocket is often referred to as the "distal" or "aryl" binding site. The proline residue, shown in light blue, binds under a flap of the protein which helps to define the S2 binding site. The location of the S2 binding site between the aryl and the guanidino binding pockets made this binding site an attractive region for the design of scaffolds for targeted small molecule libraries. The inhibitor design strategy was to identify scaffolds that would not only bind into this S2 domain but would also contain the correct number and location of sites for combinatorial explosion into both the aryl and guanidino binding sites.

Results from our initial X-ray crystallography work with a number of known, structurally distinct thrombin inhibitors gave us confidence that a suitable scaffold could be identified. Shown in Figure 2 is an overlay containing three structurally distinct series (5) derived from the D-Phe-Pro-Arg scaffold, an argatroban-like scaffold, and napsagatran (Ro-46-6240) scaffold bound into the active site of thrombin. While these compounds served as reference compounds for this work, they also provided key insights concerning inhibitor binding in the thrombin active site. Immediately evident was the fact that these inhibitors placed an aryl substituent (upper left) into the aryl binding site and its charged functionality (lower center), as anticipated, into the guanidino binding pocket. What was more interesting was the broad structural diversity observed in the "scaffold region" of these inhibitors and the multiplicity of modes used by these inhibitors to gain access into the specificity pockets of the active site. These observations led us to believe that structure-based design methods could be used to design or identify proprietary, non-peptide scaffolds that were not only complementary to the S2 binding site but also contained the appropriate number and location of sites that could be functionalized in a high throughput manner.

Early efforts in combinatorial chemistry often focused on maximizing the number and diversity of library members; first with oligomeric (10-12) and subsequently with small molecule libraries (13-15). Consequently, scaffolds for combinatorial experiments were often selected strictly based on convenience (synthetic accessibility and appropriate sites for combinatorial functionalization) irrespective of the therapeutic target(s). For example, a Kemp's triacid-based library (14) screened against thrombin produced as its most potent member SEL-2800 (K_i = 4 μM, Figure 3). The marginal performance level of this large library versus thrombin may have been due to the non-optimal nature of the scaffold.

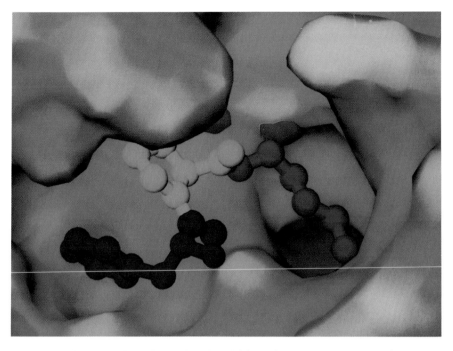

Figure 1. Thrombin active site with bound D-Phe-Pro-agmatine **1**.

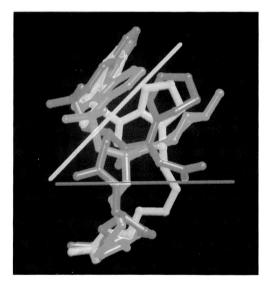

Figure 2. Structural diversity in the S2 binding region. Atoms above the yellow line bind into the "aryl" binding site. Atoms below the red line bind into the S1 or guanidino binding site.

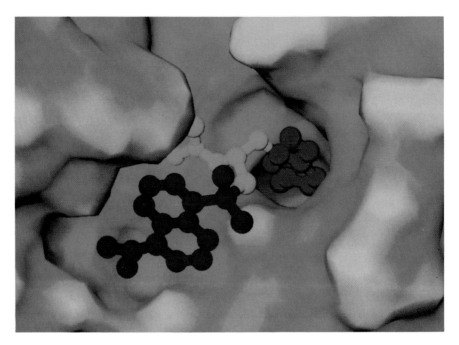

Figure 5. Thrombin-bound conformation of aminopyridine-based inhibitor **4**.

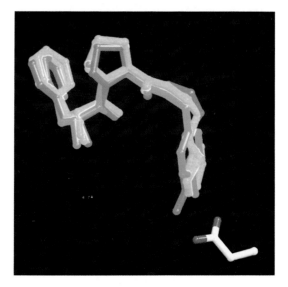

Figure 6. Overlay of the determined binding modes for inhibitors **1, 5-7**.

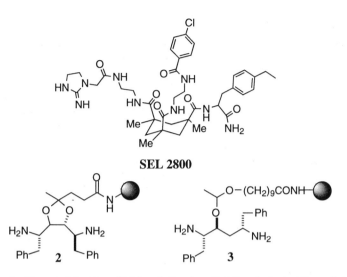

SEL 2800

Figure 3. Kemp's triacid scaffold and diamino diol **2** and alcohol **3** scaffolds.

In contrast, it is often advantageous to design scaffolds which not only present sidechains into relevant binding pockets but, just as importantly, have structural features designed into the scaffold itself which allow for key interactions between the scaffold and its biological target. For example, Abbott's targeted library of C_2-symmetric HIV protease inhibitors (*16*) successfully employed diamino diol **2** and diamino alcohol **3** templates (refer to Figure 3). The hydroxyl groups present on these scaffolds form critical hydrogen bonds to the active site and are essential for the observed potency of these series (*17*). Reports such as this point to an increasing exploitation of structural information in future scaffold design. In addition to aiding *de novo* design of scaffolds for targeted libraries, structural information is also valuable for evaluating whether any one (or more) of the ever-growing number of published scaffolds is suitable for a particular target.

When desirable, structural information about related enzymes (in this case, the structures for a variety of other proteases are known) can also be factored into library design to enhance the probability that the library will demonstrate selectivity for the desired target. Once initial library members have been synthesized and found to be active, X-ray crystallography is used to identify their binding mode(s). This information is often used to enhance the performance of that library or guide the design of subsequent libraries.

While structural information plays a central role in the design and identification of scaffolds and templates for targeted libraries, this information impacts other aspects of library design as well. The architecture of the active site can be used to predict what kinds of scaffold substituents ought to maximize binding in particular subsites. In the case of thrombin, structure suggests placement of aryl or hydrophobic residues into the aryl binding site and polar substituents into the S1 guanidino binding site. Anticipating the nature of scaffold substituents that are most complementary to particular subsites in the active site is a useful exercise early in the library design process, since the strategy for synthesis of that targeted library is often dictated by the chemical properties of these substituents.

In addition to the above mentioned uses, structural information also aids in the identification of both the appropriate location and method of attachment of the ligand to the solid phase supports. In some cases, the functionality that remains on the scaffold after cleavage can play a major role in the binding of the library molecule to the receptor. One such example is the alcohol functionality present on the diamino diol and diamino alcohol cores in Abbott's C_2-symmetric HIV protease libraries mentioned above. In other cases, the goal is to use structural information to select the point of attachment at a site where little or no negative impact on potency is anticipated and/or where the chemical properties of the remaining functional group might best modify other factors such as solubility.

Just as structural information plays a key role in the design and prioritization of ideas for targeted small molecule libraries, this data can also be used to prioritize the synthesis of library compounds based on maximal probability of interacting favorably with the target enzyme or receptor. It is clear that advances in robotic synthesis and instrumentation will soon give us the potential to synthesize many more compounds than it will be practical or affordable to analyze, screen, track, and archive. As an example, for a simple library created using amine and acid condensations onto a given amino acid scaffold, over a million compounds can be produced from commercially available reagents. In the past, the task of choosing specific members from a larger collection or set to test in a biological screen has been based on a variety of selection strategies. Many of these require little if any structural knowledge of the target receptor. Examples of selection strategies include molecular diversity (18,19), cluster analysis (20), and similarity (18,20-22) to known actives. Pharmacophore models or SAR hypotheses based on known active leads are additional examples. However, as more structural information about the target is known, more sophisticated strategies can be employed to prioritize the synthesis of molecules based on the architecture of the active site and its unique subsite properties, such as volume, hydrophobicity, and charge. A first generation example of library selection based on the unique properties of the guanidino binding site will be discussed later in this chapter. Further, the number and location of hydrogen bonding opportunities as well as distance and angular constraints between subsites are also useful selection criteria. Modified docking routines may represent an additional prioritization strategy. Rather than using docking routines (23,24) to prioritize between molecules that were "designed" to be active, these protocols could be used in a cruder yet perhaps more powerful way to screen out the vast majority of compounds in a potential library that have little if any chance of binding to a given receptor. Sophisticated selection strategies based on structural data and other relevant criteria are currently under development in numerous laboratories and are expected to enhance the efficiency of future combinatorial discovery efforts.

While much of the initial effort in combinatorial chemistry focused on the creation of large collections of diverse compounds, the most likely developments in the future are those which exploit the potential of high-throughput chemistry in smarter and more creative ways. After all, the goal of high-throughput chemistry, within the context of pharmaceutical discovery, is not just to generate large numbers of diverse compounds, but is to help identify and optimize new drug candidates in as resource- and time-efficient manner as possible. One way to exploit this emerging technology is to use structural data to focus efforts on the design of templates which bias libraries towards a particular target or target class, and to identify and prioritize the synthesis of those library members which have the greatest probability of interacting with the desired target(s). Applying the strategies discussed above has led to the development of small molecule libraries directed against thrombin. In addition to producing leads for our in-house structure-based design efforts, preliminary results suggest that these libraries can be successfully redirected toward other protease targets as well. Details of these investigations are beyond the scope of this chapter and will be reported separately elsewhere.

Impact of High-Throughput Chemistry on Structure-Based Design

We believe that the effectiveness of structure-based design can be enhanced by exploiting the speed and sampling power of combinatorial and automated chemistry. One impact of high-throughput chemistry on structure-based design will be as a rich source of novel leads and scaffolds. Crystallographic analysis of these leads with its receptor will identify new binding modes which can be further exploited using structure-based techniques. With high-throughput chemistry providing new leads, one result may be that less emphasis is placed on *de novo* design methods *(25,26)* of lead identification. Since many of the combinatorial leads will suffer from one liability or another, structure-based drug design will often be used for further property optimization and refinement. The goal of these efforts might be to further increase potency (e.g. by introducing conformational constraints), correct metabolic or stability problems, or to modify solubility or pharmacokinetic parameters.

One of the merits of automated and combinatorial chemistry is its ability to rapidly generate structure activity relationships (SAR) by virtue of the library's synthetic accessibility. However, most current structure-based discovery efforts are not working with scaffolds and templates that are amenable to current solid-phase or solution phase high-throughput techniques. Our own structure-based drug design efforts directed towards non-guanidino-based thrombin inhibitors were also impeded by this limitation. An objective of our initial work was to rapidly identify functionalities that could serve as guanidino replacements. To date, most thrombin inhibitors suffer from short duration of action and poor oral bioavailability. Many of these compounds bear highly basic, charged functional groups such as a guanidine or amidine moieties. The poor bioavialability of these inhibitors may be due in part to the presence of these basic groups *(5)*.

Several guanidino mimetics were designed using structure-based techniques and mounted on an argatroban-like scaffold. The structure of one of these (amino-pyridine **4**) is shown in Figure 4. This scaffold was chosen based on demonstrated potency *(27)* and lack of functionality capable of interacting with the catalytic triad, which is a highly conserved characteristic of serine proteases. Together, we envisioned that this choice of scaffold and guanidino mimic might enhance the selectivity profile of these inhibitors versus other serine proteases.

This aminopyridine-based inhibitor **4** was found to exhibit sub-micromolar inhibition of thrombin with moderate selectivity (Table I). Crystallographic analysis of this inhibitor bound to thrombin reveals that the guanidino mimetic binds to Asp 189 located deep within the S1 pocket of the enzyme,   and that the exocyclic primary amine

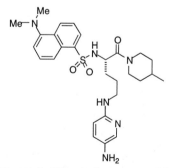

Figure 4. Chemical structure of **4**

Table I. Screening Data for 4

Assay	K_i (μM) or (% I at Screening Dose)
Thrombin	0.36
Plasmin	8.3
Urokinase	(0) [a]
FXa	39
Chymotrypsin	30
Trypsin	(0) [a]
Elastase	(0) [a]

[a] screened at 6 μM

displaces a water molecule present at the bottom of the pocket (see Figure 5).* It was quickly appreciated, however, that rapid identification of guanidino mimics employing this scaffold would be difficult since each guanidino mimic required extensive synthetic effort.

A strategic decision was made to employ the more convenient D-Phe-Pro-X dipeptide scaffold in order to facilitate the rapid identification of guanidino mimics. Using the constraints of the S1 pocket size, potential guanidine mimetics were chosen from a host of commercially available primary amines, which, in turn, were further screened by the presence of additional hydrogen-bonding and hydrogen-accepting moieties.

A library containing these 80 potential P1 mimetics was then synthesized in parallel and assayed in a high-throughput manner against a panel of serine proteases. We generated a relative ranking of these P1 mimetics of which a sampling is shown in Table II. These compounds inhibit thrombin within an order of magnitude of D-Phe-Pro-agmatine **1** which served as a reference for this library.

Table II.   Protease Inhibition of Compounds 1, 5-7

K_i (µM) or (% Inhibition at Screening Dose)

Assay	Compd	1	5	6	7
Thrombin		0.35	2.6	4.1	2.9
Plasmin		(0) [a]	(0) [f]	(0) [e]	15
Urokinase		(0) [a]	(0) [c]	(0) [d]	(0) [b]
FXa		(0) [a]	(0) [f]	(0) [e]	31
Chymotrypsin		41	110	(0) [e]	(0) [b]
Trypsin		4.3	(0) [f]	(0) [d]	19
Elastase (HLE)		(0) [a]	(0) [c]	(0) [d]	(0) [b]

[a] screened at 6 µM; [b] screened at 13 µM; [c] screened at 16 µM; [d] screened at 19 µM; [e] screened at 60 µM; [f] screened at 90 µM

An overlay of the determined binding modes for inhibitors **1, 5-7** bound to thrombin is shown in Figure 6.* The major binding energy for these compounds is derived from interaction of the mimetic at the P1 site with Asp189 at the bottom of the S1 pocket.   Additionally, these guanidino mimetics only minimally perturb the positioning of the remainder of the inhibitor (D-Phe-Pro-).   We would therefore argue that the *differences* in the potency of these and other library members are controlled by the effects of the mimetics within the S1 subsite, and does not reflect on the affinity of the scaffold.   This strategy resulted in the rapid identification of numerous additional

*NOTE: Please see color illustrations for Figure 5 and Figure 6 on page 20.

charged and uncharged guanidino mimetics (at physiological pH) and enabled a useful SAR to be developed. A full description including experimental details of this work will be reported elsewhere.

We are currently integrating this rapidly generated structural and SAR information with potent structure-based leads that are less amenable to high-throughput synthesis as outlined in the cartoon of Figure 7. X-ray crystallography plays a key role in determining the exact binding modes for lead compounds emerging from both approaches. Recombination strategies suggest new hybrid entities that incorporate the best design features or characteristics of one or both series. Crystallography plays an important role since recombination will often require redesign of either substituents and/or scaffolds. We believe that this paradigm is general and will be applicable to many targets.

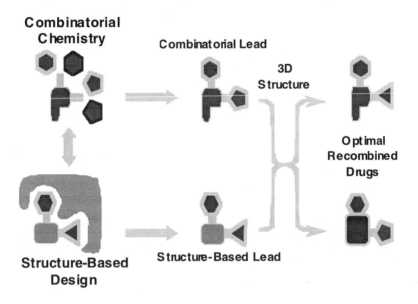

Figure 7. Synergistic discovery approach: integration of high-throughput chemistry and structure-based design.

Conclusion

The work described above demonstrates a new synergistic approach to drug design that addresses some of the individual limitations of combinatorial chemistry and structure-based design, and provides an important new paradigm for drug discovery. From a structure-based perspective, combinatorial high-throughput techniques can be used to parallelize the traditionally serial process of lead refinement, and enhance the structure-activity space that can be effectively investigated. At the same time, structural information can be employed to tailor chemical libraries directed against specific molecular targets and increase the probability of finding active hits. By combining the speed and sampling power of combinatorial chemistry with the precision of structure-based design, this approach eliminates some of the shortcomings of traditional drug discovery and provides an important new strategy for generating novel, patentable compounds in competitive areas of inhibitor design.

It is clear that in the near future a new generation of integrated drug discovery systems will emerge that will exhibit an unprecedented level of integration of diverse technologies such as high-speed automated synthesis, high-throughput screening, structure determination, informatics, and computational intelligence. To make this possible, advanced data management systems will be required to collect, store, and access the massive amounts of information produced by high-speed parallel synthesis. As a direct consequence of this information explosion, an increasing proportion of the decision making process will be delegated to intelligent machines. These machines will collect and assimilate vast amounts of raw data, form predictive models of biological activity, design follow-up experiments, and directly supervise and control the operation of a wide variety of robotic hardware. A description of the first system that achieves this level of integration, known as DirectedDiversity®, has already appeared (28). With the concomitant development of novel assays, it may soon be possible to develop SAR models to predict key pharmacokinetic and pharmacological properties such as intestinal absorption, metabolic stability, elimination, toxicity, and cross-over pharmacology. Incorporating such information as early as possible in the drug design cycle, along with the tight systems integration described above, will help reduce the aggregate failure rate of preclinical candidates and improve the efficiency and productivity of the drug discovery process.

Literature Cited

1. Longman, R. *In Vivo*, **1994**, *5*, 23 - 31.
2. Thayer, A. M. *Chemical and Engineering News* **1996**, February 12, 57 - 64.
3. Borman, S. *Chemical and Engineering News* **1996**, *Feb. 12*, 29-54.
4. Erickson, J. W.; Fesik, S. *Annual Reports in Medicinal Chemistry* **1992**, *27*, 271-289.
5. Kimball, S. D. *Blood Coagulation and Fibrinolysis* **1995**, *6*, 511 - 519.
6. Tapparelli, C.; Metternich, R.; Ehrhardt, C.; Cook, N. S. *Trends in Pharmacological Sciences* **1993**, 14, 366 - 376.
7. The solvent accessible surfaces shown in this chapter were produced using Connolly's Molecular Surface Program.
8. Bajusz, S.; Szell, E.; Barabas, E.; Bagdy, D.; Zsuzsanna, M. US Patent 4,346,078, August 24, 1982.
9. Wiley et al. have also recently described the thrombin-bound conformation of **1**. Wiley, M. R.; Chirgadze, N. Y.; Clawson, D. K.; Craft, T. J.; Gifford-Moore, D. S.; Jones, N. D.; Olkowski, J. L.; Schacht, A. L.; Weir, L. C.; Smith, G. F. *Bioorg. Med. Chem. Lett.* **1995**, *5*, 2835-2840.
10. Pavia, M. R.; Sawyer, T. K.; Moos, W. H. *Bioorg. Med. Chem. Lett.* **1993**, *3*, 387-396.
11. Furka, A. *Drug Development Research* **1995**, *36*, 1-12.
12. Gallop, M. A.; Barrett, R. W.; Dower, W. J.; Fodor, S. P. A.; Gordon, E. M. *J. Med. Chem.* **1994**, *37*, 1233-1251.
13. Gordon, E. M.; Barrett, R. W.; Dower, W. J.; Fodor, S. P. A.; Gallop, M. A. *J. Med. Chem.* **1994**, *37*, 1385-1401.
14. Seligmann, B.; Abdul-Farid, F.; Al-Obeidi, F.; Flegelova, Z.; Issakova, O.; Kocis,P.; Krchnak, V.; Lam, K.; Lebl, M.; Ostrem, J.; Safar, P.; Sepetov, N.; Stierandova, S.; Strop, P.; Wildgoose, P. *Eur. J. Med. Chem.* **1995**, *30*(Supplement), S 319 - S 335.
15. Terrett, N. K.; Gardner, M.; Gordon, D. W.; Kobylecki, R. J.; Steele, J. *Tetrahedron* **1995**, *51*, 8135-8173.
16. Wang, G. T.; Li, S.; Wideburg, N.; Kraft, G. A.; Kempf, D. J. *J. Med. Chem.* **1995**, *38*, 2995-3002.
17. Thaisrinvongs, S. *Annual Reports in Medicinal Chemistry* **1994**, *29*, 133-144.

18. Sadowski, J.; Wagener, M.; Gasteiger, J. *Angew. Chem. Int. Ed. Engl.* **1995**, *34*, 2674 - 2677.
19. Martin, E. J.; Blaney, J. M.; Siani, M. A.; Spellmeyer, D. C.; Wong, A. K.; Moos, W. H. *J. Med. Chem.* **1995**, *38*, 1431-1436.
20. Willett, P. *Similarity and Clustering in Chemical Information Systems*; John Wiley & Sons: New York, NY, 1987.
21. Maggiora, G. M.; Johnson, M. A. *Concepts and Applications of Molecular Similarity*; John Wiley & Sons: New York, NY, 1990.
22. Sheridan, R. P.; Kearsley, S. K. *J. Chem. Inf. Comput. Sci.* **1995**, *35(2)*, 310-320.
23. Kuntz, I. D.; Meng, E. C.; Shoichet, B. K. *Acc. Chem. Res.* **1994**, *27*, 117-123.
24. Judson, R. S.; Jaeger, E. P.; Treasurywala, A. M. *J. Mol. Struct. (THEOCHEM)* **1994**, *308*, 191-206.
25. Bohm, H-J. *Journal of Computer-Aided Molecular Design* **1992**, *6*, 61-78.
26. Bohm, H-J. *Journal of Computer-Aided Molecular Design* **1992**, *6*, 593-606.
27. Okamoto, S.; Hijikata-Okunomiya, A. *Methods in Enzymology* **1993**, 328.
28. Agrafiotis, D. K.; Bone, R. F.; Salemme, F. R.; Soll, R. M. A System and Method of Automatically Generating Chemical Compounds with Desired Properties. US Patent 5463564, October 31, 1995.

LIBRARY DESIGN:
SOLID-, SOLUTION-, AND LIQUID-PHASE COMBINATORIAL SYNTHESIS

Chapter 4

Phosphorus as a Scaffold: Combinatorial Libraries of Non-nucleotide Phosphoramidates

Alan F. Cook and Reza Fathi

PharmaGenics, Inc., 4 Pearl Court, Allendale, NJ 07401

Multicomponent combinatorial libraries have been prepared using phosphorus groups as scaffolds. Diols representing various elements of structural diversity have been converted into their dimethoxytrityl H-phosphonate derivatives and coupled together on solid supports to produce combinatorial libraries of phosphodiesters. Additional functionalities were also introduced by reaction of the H-phosphonate diester intermediates with a wide range of primary and secondary amines to produce libraries of phosphoramidates. The pool and divide method was used for library assembly. Libraries of over one million compounds possessing five variable positions were generated using this approach, and smaller libraries with fewer centers of diversity were also produced. Several methods for the introduction of fluorescent and radioactive labels were developed. These libraries will be used for identification of compounds which bind with high affinity to protein targets of therapeutic interest, in conjunction with COMPILE, an indirect method for identifying ligands in large mixtures of combinatorially synthesized compounds.

Oligonucleotides can be considered as the first type of phosphorus based combinatorial libraries to be prepared for drug discovery purposes. The initial demonstration of the concept of selection of specific sequences from large libraries led to a rapid expansion in the use of combinatorial techniques for selecting oligonucleotides with affinity for nucleic acid binding proteins, as well as other classes of proteins not normally associated with nucleic acid interaction (1). The use of oligonucleotide libraries was facilitated in part by the availability of automated synthesis techniques, which enabled very large libraries to be readily produced. Natural oligonucleotides have the additional advantage that selection procedures employing amplification by polymerase chain reaction can be utilized. The disadvantages of oligonucleotides, especially unmodified oligonucleotides, as

1054–7487/96/0030$15.00/0

potential drug candidates lie in their poor penetration through cell membranes together with their rapid degradation by nucleases, so that considerable effort has been devoted to the search for analogs which do not suffer from these limitations.

Phosphorus based compounds have not been widely used for combinatorial libraries outside of the oligonucleotide area, although they have several advantages. Firstly, the multivalent nature of the phosphorus atom provides opportunities for the introduction of three elements of diversity for each phosphorus atom by the formation of phosphate esters or phosphoramidates. Secondly, automated solid phase synthesis is well established and can be readily adapted to non-nucleotides. Thirdly, the charges on the phosphorus backbone can be manipulated so as to create charged or uncharged species. We have exploited this potential by preparing a wide variety of non-nucleotide synthons and coupling them together to prepare large libraries of non-nucleotide phosphorus based compounds of relatively low molecular weight (in the range of 500-1500). Libraries of this type will be used for lead discovery purposes in the search for compounds which affect biological targets of therapeutic interest.

Design of Libraries

Our overall objective was to prepare large libraries of compounds which could be assembled on a DNA synthesizer but would not possess the undesirable characteristics of oligonucleotides, such as large size, which might restrict uptake into cells, or instability in vivo. Libraries of many compounds can be constructed in one of two ways: a) repeated couplings with a relatively small set of monomers, such as for natural oligonucleotides, where a library of one million compounds can be generated by preparing decanucleotides in which each position consists of any of the four standard nucleotides, or b) using a much larger monomer set with fewer coupling steps. For the latter approach, a library of one million compounds can be assembled by using a set of sixteen monomers with only five positions of diversity, such as, for example, phosphoramidate libraries of general structure **1** (Figure 1). In this approach the structural elements A, B and C are derived from diols and elements D and E are derived from various amines. We have used this latter approach since it produces materials of much smaller molecular weight, a desirable feature for eventual drug development. Kim et al. have reported that the vast majority of drugs on the market have molecular weights of less than one thousand (*2*).

Selection of Synthons

The starting materials for these libraries employed a wide range of functionalized and unfunctionalized diols, most of which were commercially available, or which could readily be prepared by a short synthetic route. These are shown in Figure 2. Synthons were selected so as to provide a range of functionalities, such as hydrogen bond donors and acceptors (structures **3, 4, 5, 7, 9**), aliphatic (structure **2**) and aromatic groups (structures **3, 5-8**), and charged species (structures **3, 4, 7, 9**). Synthons with steric bulk, which might add an element of constraint to some members of the libraries, were also included (structures **8, 9**). The primary requirements were that the hydroxyl groups were primary or secondary but not

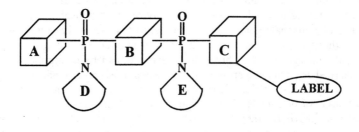

1

Figure 1. Example of a phosphorus based combinatorial library. **A**, **B** and **C** are derived from diols and **D** and **E** are derived from amines.

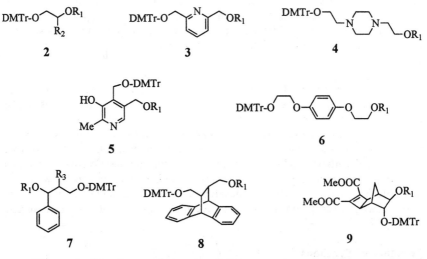

Figure 2. Structures of representative monomers used to prepare combinatorial libraries. DMTr = 4,4'-dimethoxytrityl, R_1 = HP(O)O$^-$, R_2 = H, Me, CH=CH$_2$, CH$_2$NEt$_2$ or m-CH$_2$OC$_6$H$_5$OCH$_3$, R_3 = H or NHCOCF$_3$.

tertiary, since preliminary experiments demonstrated that diols possessing tertiary hydroxyl groups were too unreactive for efficient coupling using a DNA synthesizer. Another consideration in the choice of synthons for the libraries was to select monomers which would provide a great deal of variation to the backbone (structures **3, 4, 6**), thus increasing the array of shapes in the library. Combinatorial libraries of oligomers with regular backbones which have previously been reported include peptides, oligonucleotides and peptoids *(3-6)*. Smaller libraries of phosphodiesters containing nucleotides along with non-nucleotide synthons, such as derivatized ethylene glycol phosphodiesters similar to **2**, have also been reported *(7,8)* and others could presumably be assembled from available synthons.

Coupling Methods

Two basic methods, the phosphoramidite method and the H-phosphonate method, have been most widely used for coupling of phosphorus based synthons, such as oligonucleotides. The former has found wide acceptance in the field of oligonucleotide synthesis, primarily due to the higher coupling yields and the availability of a wide array of reagents. A disadvantage of the phosphoramidite method is that it does not permit modifications to the phosphorus backbone, other than phosphorothioates by oxidation with a sulfur based reagent *(9)*. Although the H-phosphonate method was found to give slightly lower coupling yields, oligonucleotide H-phosphonate diesters have been shown to be versatile intermediates which could react with a wide variety of amines to produce phosphoramidates *(10)*. This is especially valuable for the preparation of phosphorus based combinatorial libraries, since it presents an opportunity to introduce additional sites of diversity by formation of phosphoramidates from H-phosphonate intermediates.

The individual diols were converted into their dimethoxytrityl derivatives and then treated with phosphorus trichloride to produce dimethoxytrityl H-phosphonates **2-9** using a method which has been used widely in the oligonucleotide area *(11)*. Trial couplings were then performed on a DNA synthesizer to determine the conditions required for efficient coupling of each monomer, using a commercially available thymidine-derivatized solid support for the test system. A modified H-phosphonate coupling cycle with an extended coupling time of 5 minutes was found to be satisfactory for all monomers, and trityl assays of coupling efficiency were generally in the range of 90-98%. Coupling yields of greater than 90% were considered to be satisfactory for use in the synthesis of these libraries, since the objective was to employ only 2-4 couplings so as to keep the molecular weights to a minimum. These relatively low yields would not, of course, be acceptable for the synthesis of longer sequences.

Assembly of Libraries

A. Oligomeric Phosphodiesters. Libraries of oligomeric phosphodiesters were assembled by coupling of monomer H-phosphonates to solid supports using the pool and divide method to redistribute the support between reactions *(12)*. After each coupling step, the supports were removed from the columns, suspended in acetonitrile and redistributed into the original columns using a liquid handling system and a manifold attached to a vacuum source. Reusable twist lock columns (obtained from Glen Research, Stirling, VA) were used to facilitate the process of pooling and dividing the supports. At the conclusion of the synthesis, the H-phosphonate groups were oxidized to phosphodiesters using standard aqueous iodine conditions and the labelling group was attached as the final step (see below). Using sixteen monomers and five variable elements in this way permitted the synthesis of libraries of non-nucleotide phosphodiesters with theoretically one million compounds ($16^5 = 1,048,576$) of general structure 10, as shown in Figure 3.

B. Phosphoramidates. Libraries of oligomeric phosphodiesters use only two valencies of each phosphorus atom, since the third valency is occupied with the anionic oxygen substituent. A second approach to library construction utilized three valencies of the phosphoryl group by introducing additional diversity via phosphoramidate groups. One of the advantages of this approach lies in the availability of large numbers of amines which can be directly utilized as synthons to introduce the additional elements of diversity. Additionally, the same number of compounds in the library can be produced with compounds of lower molecular weight, since the phosphorus atoms which act as scaffolds are more effectively utilized, and fewer phosphorus groups are required.

The procedure for assembly of libraries of phosphoramidates involved coupling of the H-phosphonate monomers on the DNA synthesizer, followed by the standard pool and divide step to redistribute the support. Each support was then treated with a solution of an amine in carbon tetrachloride to convert the H-phosphonate linkages to phosphoramidates. This step could be carried out either on the DNA synthesizer or manually using a syringe for the reaction vessel, although the former method was preferred. A wide range of primary and secondary amines could be used for this purpose and the following were selected for library 11: 2-(2-aminoethyl)pyridine, 1-(2-aminoethyl)piperidine, 2-(3,4-dimethoxyphenyl)ethylamine, 4-(2-aminoethyl)-morpholine, 1-(3-aminopropyl)-2-pyrrolidinone, 2-(3-chlorophenyl)ethylamine, N-isopropyl-1-piperazine acetamide, N-(3-trifluoromethyl)-phenylpiperazine, 3,3-diphenylpropylamine, thiomorpholine, 1-(2-pyridyl)piperazine, hexamethyleneimine, cis-2,6-dimethyl-morpholine, 1-(4-fluorophenyl) piperazine, 4-fluorophenethylamine and 3,5-dimethylpiperidine. Additional amines were used for libraries 12 and 13 (a total of 24 for library 12 and 36 for library 13). Trial reactions indicated that for model compounds the efficiency of amine addition to the H-phosphonate intermediate was very high. Froehler has previously shown that oligonucleotide H-phosphonates can be converted into a wide variety of phosphoramidates in high yield using a solution of an amine in carbon tetrachloride (*10*).

Amine addition was followed by a second pool and divide step, and this cycle of coupling followed by amine addition was repeated as necessary to produce an array of combinatorial libraries with structures 11-13 (Figure 4). With these libraries the phosphoramidate groups were present as a mixture of stereoisomers which were not separated. The libraries were then labelled with either a fluorescent or radioactive label using one of the methods described below. Details of the phosphodiester and phosphoramidate libraries are summarized in Table I.

$$R_1\text{-p-}M_5\text{-p-}M_4\text{-p-}M_3\text{-p-}M_2\text{-p-}M_1\text{-p-dT}$$

10

Figure 3. Structures of phosphodiester-based combinatorial libraries, where p represents a phosphodiester linkage and R_1 is either a) thymid-3'-yl, b) 5,(6)-carboxamidohexylfluorescein, or c) m-CH_3-$C^3H(OH)$-C_6H_5 .

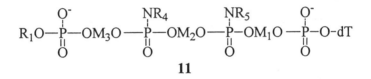

11

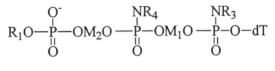

12

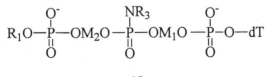

13

Figure 4. Types of libraries of phosphoramidates prepared, where R_1 is either a) ^{32}P-5'-phosphoryl-thymid-3-yl, b) 5,(6)-carboxamido-hexylfluorescein, or c) m-CH_3-$C^3H(OH)$-C_6H_5. M_{1-3} and NR_{3-5} represent sites of diversity derived from diols and amines, respectively.

Table I. Summary of the Libraries 10-13.

Structure	Variable Sites	Diols	Amines	Compounds
10	5	16	0	1,048,600
11	5	11	16	1,362,900
12	4	15	24	518,400
13	3	20	36	28,800

Labelling Methods

The libraries described above could in theory be used without labelling using certain deconvolution strategies, such as those based on activity analysis of pools (3). For our purposes however, labelled libraries were required for deconvolution using the COMPILE procedure (13), so that several alternatives for labelling were therefore explored. Two categories of label were investigated, a) fluorescent and b) radioactive. Fluorescein was selected as the fluorescent label, since it possesses a high fluorescence intensity, and reagents which can be used directly on the synthesizer are commercially available. The library attached to the solid support was thus subjected to coupling with a commercially available fluorescein phosphoramidite, and double coupling with an extended coupling time of 5 minutes each was used to drive the addition of fluorescein to completion. The library was then cleaved from the support, lyophilized and used directly for lead discovery purposes.

Radioactive methods for labelling included ^{32}P and tritium. For ^{32}P labelling, a thymidine residue was added as the last cycle on the synthesizer, and after isolation the library was labelled enzymatically using polynucleotide kinase with γ-^{32}P-labelled ATP as the donor source. The enzyme requires the presence of a terminal nucleotide but can apparently tolerate a number of modifications at the 3'-position. Since the libraries possessed several negative charges, they could be purified on a polyacrylamide gel.

Tritium was also used as a label for libraries of non-nucleotide phosphoramidates. This approach involved introduction of a labelling group possessing a ketone functionality which was subsequently reduced using tritium labelled sodium borohydride. This type of reaction was chosen because ketone functionalities can be readily reduced and sodium borohydride can be obtained with high specific activity. The meta-substituted acetophenone moiety was chosen as the labelling group, since the reduced product is unlikely to undergo decomposition under alkaline conditions via a cyclic phosphate intermediate due to the rigidity of the aromatic ring. After assembly of the library, a phosphoramidite derivative of meta-hydroxyacetophenone was used to introduce the labelling group to give a library of compounds of general structure **14** (Figure 5). The ketone functionality could be then reduced with tritium labelled sodium borohydride of high specific activity to produce tritium labelled libraries of structure **15**.

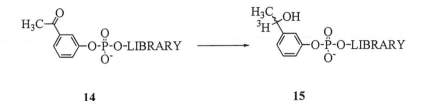

14   **15**

Figure 5. Method for tritium labelling of combinatorial libraries by reduction of the meta-hydroxyacetophenone moiety.

Capping Groups

For convenience, the initial libraries were prepared using a conventional thymidine derivatized solid support, which after cleavage from the support results in each member of the library having a thymidine group attached to its terminus. This thymidine group plays an important role in stabilizing the libraries to degradation by the concentrated ammonia conditions required for cleavage from the support. In the absence of a capping group, compounds with flexible terminal groups might be expected to undergo degradation by attack of the terminal hydroxyl group on the neighboring phosphoryl moiety to produce a cyclic phosphate intermediate followed by elimination. Degradation of this type has been demonstrated with oligonucleotides which terminated in ethylene glycol units (*14*).

  Methods have also been developed for the elimination of the terminal thymidine groups while maintaining the stability of the libraries towards alkaline degradation. The approach was to prepare a support which, when treated with ammonia, generates an amide end group, and is similar to a reported method for generating 3'-end-capped oligonucleotides (*15*). Amino-derivatized controlled pore glass was treated with O-dimethoxytrityl glycolic acid (**16**, Figure 6) to give the solid supported amide derivative **17,** which was detritylated and coupled to a second molecule of **16** to give the diglycolic ester derivative **18**. This support was used as a replacement for the thymidine derivatized support for the synthesis of libraries. After completion of the synthesis, the library was cleaved from the support by reaction with ammonia, which cleaved the ester group to produce compounds which terminated with glycolic amide functionalities (**19**). Removal of the thymidine groups and substitution with glycolic amide groups in this way results in a reduction in molecular weight of approximately 170 for each library member. In addition, this glycolic acid derivatized support can be used to introduce another element of diversity by cleavage using a variety of primary or secondary amines to produce substituted amides; this will be discussed in detail elsewhere. To our knowledge, the concept of using the linkage site to introduce an element of diversity has not been previously explored in the construction of libraries.

Summary

Phosphorus groups provide a convenient way of preparing combinatorial libraries of compounds using automated synthesizers. Diversity can be achieved from large

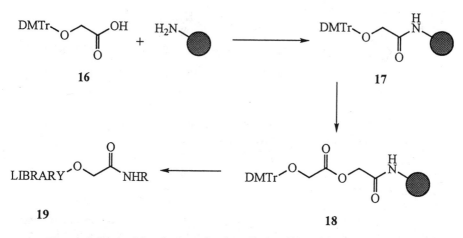

Figure 6. Method for the introduction of glycolic amide capping groups into libraries. R = H or a variety of substituted or unsubstituted aliphatic or aromatic groups.

numbers of diols and amines which are used as the starting materials for library construction. With five sites of diversity, libraries of over one million compounds can be generated, and smaller versions with fewer sites of diversity can also be prepared. These libraries are being used in the search for compounds which bind with high affinity to biological targets of interest.

Literature Cited

1. Trotta, P. P.; Beutel, B. A.; Sherman, M. I. *Med. Res. Revs.* **1995**, *15*, 277-298.
2. Kim, E. E.; Baker, C. T.; Dwyer, M. D.; Murcko, M. A.; Rao, B. G.; Tung, R. D.; Navia, M. A. *J. Am. Chem. Soc.* **1995**, *117*, 1181-1182.
3. Dooley, C. T.; Chung, N. N.; Schiller, P. W.; Houghten, R. A. *Proc. Natl. Acad. Sci. USA* **1993**, *90*, 10811-10815.
4. Ellington, A. D.; Szostak, J. W. *Nature* **1990**, *346*, 818.
5. Tuerk, C.; Gold, L. *Science*, **1990**, *249*, 505-510.
6. Zuckermann, R. N.; Martin, E. J.; Spellmeyer, D. C.; Stauber, G. B.; Shoemaker, K. R.; Kerr, J. M.; Figliozzi, G. M.; Goff, D. A.; Siani, M. A.; Simon, R. J.; Banville, S. C.; Brown, E. G.; Wang, L.; Richter, L. S.; Moos, W. H. *J. Med. Chem.* **1994**, *37*, 2678-2685.
7. Davis, P. W.; Vickers, T. A.; Wilson-Lingardo, L.; Wyatt, J. R.; Guinosso, C. J.; Sanghvi, Y. S.; DeBaets, E. A.; Acevedo, O. L.; Cook, P. D.; Ecker, D. J. *J. Med. Chem.*, **1995**, *38*, 4363-4366.
8. Hébert, N.; Davis, P. W.; DeBaets, E. L.; Acevedo, O. L. *Tetrahedron Letts.* **1994**, *35*, 9509-9512.
9. Iyer, R. P.; Phillips, L. R.; Egan, W.; Regan, J. B.; Beaucage, S. L. *J. Org. Chem.* **1990**, *55*, 4693-4699.

4. COOK & FATHI *Phosphones as a Scaffold* **39**

10. Froehler, B. C. *Tet. Letts.* **1986**, *27*, 5575-5578.
11. Ott, G.; Arnold, L.; Smrt, J.; Sobkowski, M.; Limmer, S.; Hofmann, H.-P.; Sprinzl, M. *Nucleosides & Nucleotides* **1994**, *13*, 1069-1085.
12. Furka, A.; Sebestyen, F.; Asgedom, M.; Dibo, G. *Intl. J. Pept. Prot. Res.* **1991**, *37*, 487.
13. Beutel, B. A.; Bellomy, G. R.; Waldron, J. A.; Voorbach, M. J.; Bertelsen, A. H., manuscript submitted.
14. Fontanel, M.-L.; Bazin, H.; Teoule, R. *Nucleic. Acids Res.* **1994**, *22*, 2022-2027.
15. Hovinen, J.; Guzaev, A.; Azhayev, A.; Lönnberg, H. *Tetrahedron* **1994**, *50*, 7203-7218.

Chapter 5

Convergent Solid-Phase Synthesis of Phosphoramidate Combinatorial Libraries

Normand Hébert

Isis Pharmaceuticals, 2292 Faraday Avenue, Carlsbad, CA 92008

A strategy for preparing combinatorial libraries of phosphoramidates on solid support is described. Protected aminodiol scaffolds are linked to a solid support, and functionalized with diverse reagents. H-phosphonate coupling of a second scaffold followed by oxidation in the presence of primary and secondary amines provides phosphoramidates. A comparison of different deconvolution results is also provided.

The pharmaceutical industry relies on several sources of new chemical compounds in its search for pharmacologically useful products. Historically, natural products screening, chemical industry products and medicinal chemistry have provided compounds to be developed as drugs. By systematic exploration of chemical structure space around a lead molecule compounds move from the research lab into the development pipeline. This process can be likened to a search of chemical diversity space from a gross (lead discovery) to a fine level (SAR), resulting in the optimization of pharmacological properties. The accessible chemical diversity is largely uncontrolled in natural products screening, and fine tuning is limited by synthetic difficulties. A lead from a historical chemical collection may be a unique compound or a member of a Structure Activity series prepared for an entirely different target. A chemical collection is also a static and finite resource, since the supply of compounds is limited and new analogs are not readily available. Biological extracts are also limited for several reasons. First, if an active compound is found its structure must be determined by time consuming spectroscopic means or by total synthesis. Secondly, the amount of material in a sample may be vanishingly small, and the supply uncertain. Third, the same active structures are being found over and again from different sources. Combinatorial chemistry methods are rapidly gaining in importance because they provide a means of circumventing these limitations (1,2). Either through parallel or simultaneous synthetic methods, large numbers of compounds can be made, and a systematic search of chemical property space can be achieved.

1054–7487/96/0040$15.00/0

A combinatorial drug discovery program can be used in two complementary ways: lead discovery and lead optimization. The discovery of a new chemical with a novel biological activity is the more difficult of the two processes (*3*). New proteins are continuously being identified through genomic sequencing for which detailed structures are not available. In many cases the ligands themselves are not well characterized, and functional assays are the only means of measuring biological activity. Combinatorial chemistry provides significant advantages over natural products or historical collections in these cases.

In the following sections our approach to the creation of combinatorial libraries which cover chemical diversity space in a controlled manner will be described. Our methods seek to create unbiased diverse compounds which search chemical space not yet examined by the pharmaceutical industry. We have also been studying the effectiveness of different deconvolution methods to identify the active components in combinatorial libraries, and a comparison of theoretical and experimental results will be discussed.

Scaffolds

Numerous groups have studied the properties of organic compounds which contribute to their biological activity. In general terms, certain portions of a drug molecule interact with the biological effector in a specific manner by an alignment of the electrostatic, Van der Waals, hydrophobic and dipolar interactions. In the absence of any covalent interactions between the drug and the receptor, the molecule must "fit" into an appropriate surface on the target with sufficient affinity to interfere with the other binding interactions of the target. Different effectors of a biological response can be compared to infer the specific functional groups and their 3-dimensional arrangement responsible for the activity. This information defines a pharmacophore. In designing a combinatorial library it is highly desirable to include fragments with highly diverse chemical properties. This ability to include many different types of functional groups is most useful if it is also possible to create different spatial arrangements, or footprints. Our goal has been to devise libraries which allow the incorporation of a wide variety of functionalities and substituents into a structurally diverse set of scaffolds, each creating a set of shapes which define different pharmacophores. Thus the physicochemical properties of the substituents can be varied independently of the scaffold footprint, rigidity and the relative orientation of the functional groups. This allows for an extensive search of diversity space within a window of physicochemical properties.

A tactical decision is the requirement that the library synthesis be convergent (*4*). In practice this means that the synthetic route allows functional groups to be added to the scaffold moiety during the library synthesis rather than by assembling individual functionalized monomers. This has implications on the ease of library synthesis. Thus the addition of a new functional group to a library is done with a new reagent, rather than requiring the multi-step preparation of a new monomer. The caveat is that functional groups must be chosen based on reagent availability. The advantages of using off the shelf reagents to prepare libraries are significant. The time can be spent preparing libraries and developing methodology rather than maintaining a supply of synthesis reagents. In principle, a universal library or a biased library designed to give SAR information (a library of analogues around a lead) can be generated rapidly by choosing either diverse or similar reagents.

We have developed a method for the convergent attachment of functional groups to a scaffold monomers on solid support. This allows the use of structurally diverse, protected aminodiols for library synthesis without the need to synthesize scores of monomers. The obvious choice of protecting group for the amine is an Fmoc group. This group survives mildly basic conditions, and is readily removed by treatment with piperidine in DMF. Typical scaffolds are shown (Figure 1). The amino group is protected as the Fmoc derivative, and one of the two hydroxyls is converted to the DMT ether. This allows orthogonal deprotection of either functionality for further elaboration. The remaining free OH can then either be converted to the H-phosphonate monoester (5), or used as a handle for attachment to solid support. In designing our scaffolds we sought to create a variety of shapes and degrees of rigidity, while maintaining synthetic accessibility. For example, the products derived from cyclic versus linear scaffolds will be very different in terms of the conformations accessible and the energetic differences between them. A comparison of the different scaffolds shows marked differences in flexibility, and in the positioning of the functional groups in space, as well as the number of families of related structures for each scaffold.

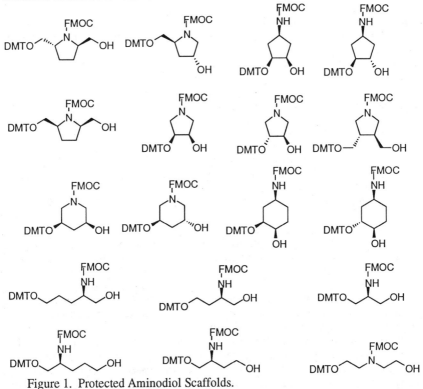

Figure 1. Protected Aminodiol Scaffolds.

Numerous methods exist for linking compounds to a solid support. Most of these have been developed for peptide or oligonucleotide synthesis. The former are methods which, for the most part require the use of relatively acidic trifluoroacetic acid solutions for cleavage from the solid support. In oligonucleotide synthesis an ester linker and base labile protecting groups are preferred due to the sensitivity of DNA, particularly purines, to acidic conditions. We preferred to link the scaffolds to a Tentagel resin via a succinate linkage (6) (Figure 2).

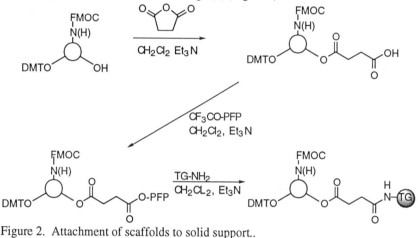

Figure 2. Attachment of scaffolds to solid support..

We established the stability of the ester to all the necessary synthetic conditions, including Fmoc removal and phosphoramidate synthesis. In conjunction with the use of the succinate linker, the DMT protecting group on the scaffold can be removed under mild acidic conditions (3% Trichloroacetic acid in CH_2Cl_2. Other reactive functional groups can be protected during library synthesis as BOC derivatives which are stable to these conditions indefinitely. Deprotection of the library can therefore be done on the solid support by acidolysis with the stronger trifluoroacetic acid, followed by cleavage of the library from the resin using aqueous ammonia. A wide variety of substituents can be introduced by acylation, sulfonylation or alkylation of the scaffold on solid support (Figure 3).

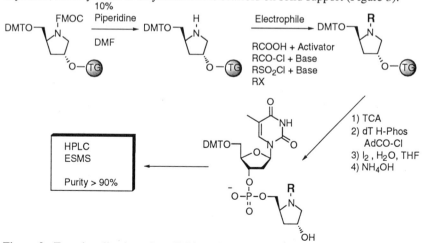

Figure 3. Functionalization of scaffolds and quality control.

A wide range of commercially available reagents can be introduced into combinatorial libraries using these three methods. All of the reagents are tested for suitability for use in a combinatorial library using a model system which uses all the preparative conditions. Characterization of the products is done by electrospray mass spectrometry (ESMS) and by HPLC. Dozens of reagents have already been found to be suitable for library synthesis in this way.

Another consideration in combinatorial library design involves the method of linking different scaffolds together. Several examples currently exist in the literature. Peptide libraries can be considered as an example: the amino acids are linked via amide bonds to generate oligopeptides. Peptoid libraries (oligo N-substituted glycines) have also been made by coupling protected N-substituted glycines (7), although a more elegant route is now preferred. Libraries of carbamates (8) and ureas (9) have also been described. Phosphodiester libraries have been made by coupling phosphoramidite monomers (10). In these examples the scaffolds must all be prepared individually as monomers, and there is little or no possibility of utilizing the linkage itself to introduce additional diversity. An exception to this is the peralkylation of the amide NH of oligopeptides on solid support by the Houghten group (11). The libraries produced in this way only incorporated a single substitution at all amide bonds in all library molecules. In the peptoid example a single scaffold is used, but the synthetic route allows each substituent to be added to a precursor on solid support without the need to prepare each monomer.

We became interested in extending the diversity of our libraries by taking advantage of the chemistry of phosphonates. It was recognized that an alternative coupling methodology involving H-phosphonate chemistry had a number of advantages (Figure 4).

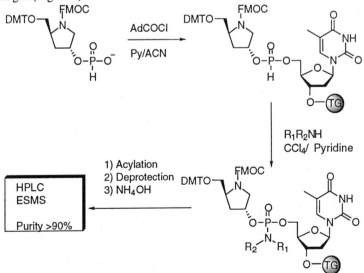

Figure 4. Synthesis of phosphoramidates and quality control.

The preparation of H-phosphonate monoesters and their coupling via mixed anhydrides proceeds in high yield under standard conditions. In addition, the H-phosphonate diester produced can be oxidized under different conditions to give phosphodiesters, phosphorothioates, and phosphoramidates (*12*). In the latter case, the H-phosphonate diester intermediate undergoes an Arbuzov type reaction with carbon tetrachloride to give a phosphoryl chloride, which can be trapped with a wide range of primary and secondary amines.This methodology is ideally suited for the preparation of combinatorial libraries since diverse functionalities can be introduced at the linkage between two scaffolds.

The synthesis of phosphoramidates is fully compatible with the functionalization of the scaffold as shown. Removal of the Fmoc protecting group followed by acylation, deprotection and cleavage from the solid support gives results which are completely comparable to those obtained in Figure 3. Moreover, the results obtained apply equally to any of the scaffolds depicted in Figure 1. Figure 5 depicts the complete scheme for the synthesis of phosphoramidate combinatorial libraries. In essence, only two components must be prepared by traditional synthetic methods: the scaffolds on solid support , and the corresponding H-phosphonates. The scaffolds serve as the structural elements of the library components, and the diverse functional groups are introduced by solid phase synthesis using an automated synthesizer (*13*).

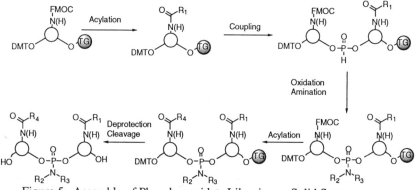

Figure 5. Assembly of Phosphoramidate Libraries on Solid Support.

Comparison of Pooling Strategies.

One frequently asked question regarding the screening of mixtures of compounds is whether the most active molecule in the library has been found. The question arises from the fear that the complexity of a mixture will cause a useful drug lead to be missed in the screening process. It is undeniable that bioactive compounds have been isolated from plant and micro-organism extracts, but the possibility exists that actives will be missed in mixtures which may contain up to millions of compounds. The corollary question is, what is the best way to discover active molecules from these mixtures. Obviously, if compounds are screened one at a time the activity of each can be determined with certainty, although this method may be the least efficient in terms of resources and time. If mixtures are screened, how are the compounds pooled in order to be reasonably sure of success. These are some of the questions we have been addressing theoretically and experimentally. Using an RNA hybridization example, a model of library deconvolution was developed (*14*.) The energy of interaction between target RNAs of different lengths and all possible

9-mer hybridization partners was calculated. This model was used to evaluate the performance of iterative deconvolution methods. Among the variables which were examined were the order of deconvolution, the effect of experimental error, and the pooling strategy(*15*). We tested the validity of the calculated results using a phosphodiester library composed of 729 trimers and 81 dimers from 9 phosphoramidite monomers (*16*) (Figure 6) which had previously been shown to contain inhibitors of PLA$_2$. (*17*)

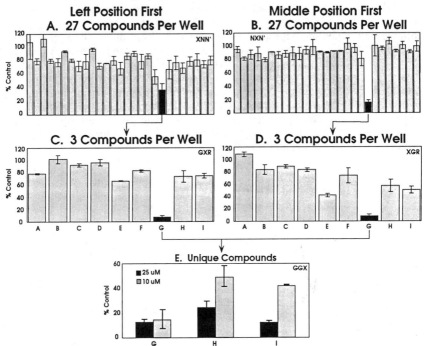

Figure 6. Deconvolution of Phosphodiester Library: inhibition of PLA$_2$. Panel A :Effect of unrandomization order for pooling by fixed position. Molecules in each XNN' subset contained a single monomer in the first position, a mix of all nine monomers in the second position and a mix of three monomers) in the third position. The NXN' subsets (panel B) were similar except position 2 contained a fixed monomer and position 1 was a mix of all nine monomers. In the second round, the 27 molecules from the most active round 1 subset were divided into nine subsets with three compounds per subsets. Molecules in each GXR subset (panel C) contained G in position 1, a single monomer in position 2 and a mix of three monomers in the third position. Molecules in each XGR subset (panel D) contained a single fixed monomer in position 1, G in position 2 and a mix of three monomers in the third position. In the third round (panel E), the three compounds in GGR were synthesized and tested as unique compounds. Adapted from (*16*).

The pooling strategy is a description of the composition of the mixtures which are to be screened. For example, in an iterative deconvolution by fixed position, all the molecules with a certain functionality at a given position are grouped together. It is also possible to pool a library randomly by mixing individual compounds (Figure 7). In both methods there are many ways in which the compounds can be mixed.

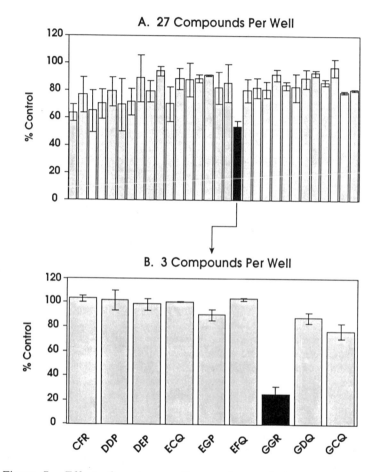

Figure 7.   Effect of random pooling on deconvolution. A) The 810 compounds were randomly pooled into 30 subsets of 27 compounds per sample. B) The most active round 1 subset was partially unrandomized in round 2 to nine pools containing three compounds per pool. Adapted from (*16*).

The calculations show that there is very little difference in the success rate with any of these strategies, albeit slightly better results are obtained if the best molecules are kept together in the early rounds. This result is borne out experimentally: when the phosphodiester library was pooled randomly, the same active compound was identified as in the iterative deconvolution above. Position scanning performed worse in the model than did iterative deconvolution. In the hybridization model, there is a problem of register complexity which means that several deconvolution pathways can lead to molecules of equal activity, and a single alignment is not found. In practice, position scanning identifies active compounds because there is frequently one interaction which is dominant in the deconvolution, although the most active compound is rarely found by a single screening experiment (16).

Furthermore, contrived pooling strategies can be used: "hard" pooling is the situation where the best molecule is alone in a pool with inactive molecules, and all the next best are grouped together in the next subset. This corresponds to the worst case scenario, which can only be achieved if the activity of the individual compounds is known beforehand. As expected this pooling strategy fails to identify the most active compound. Any combinatorial library is the result of a series of synthetic steps carried out either on individual compounds in parallel or on a mixture of starting materials. In particular, split-bead synthesis is a convenient method for the synthesis of large numbers of compounds. Any combination of fixed and random positions can be obtained in a library of compounds by the appropriate sequence of mixing and splitting operations. There is however a synthetic advantage to fixing the last position in the library. The theoretical model shows that there is very little influence of the order of unrandomization on the likelihood of finding a molecule with good activity. Experimentally it was found that fixing either the left or the middle position in the trimer led to the same final compound, although the discrimination was easier while fixing the middle position in this case. The conclusions which can be drawn from these studies is that iterative deconvolution by fixed position is successful in discovering actives, and has synthetic advantages over other pooling strategies. The pool size does not adversely affect the likelihood of discovering the active compound, but may make the identification of the active pool easier by facilitating the discrimination between pools. In practical terms the number of pools synthesized is determined mostly by the synthetic capacity. It is a balance between the number of samples which can be screened and the number of pools which can be handled by the chemist. With the advent of automation and parallel synthesis methods, this latter number is rapidly increasing.

Conclusion

In this paper, some of the considerations which govern the design of combinatorial libraries in our group were outlined. It should be understood that many of these concepts are broadly applicable, and that one of our goals is to use different chemical methods which follow these guidelines. The study of deconvolution has helped shed some light on our screening approaches, and can give valuable insight into the nature of molecular landscapes in the context of drug discovery. Combinatorial chemistry is now becoming generally accepted as a useful method for drug discovery, and the tactics and strategies are perhaps as diverse as the libraries themselves.

References

1. Gordon, E. M.; Barrett, R. W.; Dower, W. J.; Fodor, S. P. A.; Gallop, M. A., *J. Med. Chem.,* **1994**, *37*, 1385.
2. Terrett, N. K.; Gardner, M.; Gordon, D. W.; Kobylecki, R. J.; Steele, J., Tetrahedron, **1995**, 51, 8135.
3. Ecker, D.J., Crooke, S.T., *BioTechnology*, **1995**, *13*, 351.
4. Zuckermann, R. N.; Kerr, J. M.; Kent, S. B. H.; Moos, W. H., *J. Am. Chem. Soc.* **1992**, *114*, 10646.
5. Froehler, B. C.; Matteucci, M. D.; *Tetrahedron Lett.* **1986**, *27*, 469.
6. Lukhtanov, E. A.; Kutyavin, I. V.; Gamper, H. B.; Meyer, R. B. Jr., *Bioconjugate Chem.*, **1995**, *6*, 418.
7. Simon, R. J.; Kania, R. S.; Zuckermann, R. N.; Huebner, V. D.; Jewell, D. A.; Banville,S.; Ng, S.; Wang, L.; Rosenberg, S.; Marlowe, C. K.; Spellmeyer, D. C.; Tan, R.; Frankel, A. D.; Santi, D. V.; Cohen, F. E.; Bartlett, P. A., *Proc. Natl. Acad. Sci. USA*, **1992**, *89*, 9367.
8. Cho, C. Y.; Moran, E. J.; Cherry, S. R.; Stephans, J. C.; Fodor, S. P. A.; Adams, C. L.; Sundaram, A.; Jacobs, J. W.; Schultz, P. G., *Science*, **1993**, *261*, 1303.
9. Burgess, K.; Linthicum, D. S.; Shin, H., *Angew. Chem. Int. Ed. Engl.* **1995**, *34*, 907.
10. Hebert, N., Davis, P. W., DeBaets, E. L., Acevedo, O. L., *Tetrahedron Lett.* **1994**, *35*, 9509.
11. Ostresh, J. M.; Husar, G. M.; Blondelle, S. E.; Dörner, B.; Weber, P. A.; Houghten, R. A., *Proc. Natl. Acad. Sci. USA*, **1994**, *91*, 11138.
12. Froehler, B. C., *Tetrahedron Lett.* **1986**, *27*, 5575.
13. Lakshari, D. A.; Hunicke-Smith, S. P.; Norgren, R. M.; Davis, R. W.; Brennan, T., *Proc. Natl. Acad. Sci. USA*, **1995**, *92*, 7912.
14. Freier, S. M.; Konings, D. A. M.; Wyatt, J. R.; Ecker, D. J., *J. Med. Chem.* **1995**, *38*, 344.
15. Konings, D. A. M.; Wyatt, J. R.; Ecker, D. J.; Freier, S. M., *J. Med. Chem.* submitted.
16. Wilson-Lingardo, L.; Davis, P. W.; Ecker, D. J.; Hébert, N.; Acevedo, O.; Sprankle, K.; Brennan, T.; Schwarcz, L.; Freier, S. M.; Wyatt, J. R., *J. Med. Chem.* submitted.
17. Davis, P.W., Vickers, T.A., Wilson-Lingardo, L., Wyatt, J.R., Guinosso, C.J., Sanghvi, Y.S., DeBaets, E.A., Acevedo, O.L., Cook, P.D., Ecker, D.J., *J. Med. Chem.*, **1995**, *38*, 4363.

Chapter 6

Solid-Phase Synthesis, Characterization, and Screening of a 43,000-Compound Tetrahydroisoquinoline Combinatorial Library

Michael C. Griffith[1], Colette T. Dooley, Richard A. Houghten[1], and John S. Kiely

Torrey Pines Institute for Molecular Studies,
3550 General Atomics Court, San Diego, CA 92121

A library of 43,000 tri-substituted tetrahydroisoquinolines was synthesized on solid support using a three step procedure. The resulting library was screened for activity radioreceptor-binding assays for mu and kappa opioid receptors and for the sigma receptor-binding assay. Screening as pools of 836 compounds demonstrated activity in all three assays.

Combinatorial libraries are currently under intensive use as sources of new preclinical lead candidates as well as for the rapid optimization of existing lead compounds (1). Combinatorial synthetic chemistry continues to advance as the concepts first successfully illustrated with libraries of peptides (2, 3, 4) are increasingly applied to other classes of molecules. These libraries of small organic compounds can be either linear (peptidomimetics (4), peptoids (5), polyamines (6), beta-mercapto ketones (7)) or cyclic (benzodiazepines (8, 9), thiazolidinones (10), hydantoins (8)). Goff et al recently reported a solid-phase isoquinoline synthesis based on the intramolecular Heck cyclization of a peptoid (11). Here we describe the synthesis of a large combinatorial library of trisubstituted tetrahydroisoquinolines from resin-bound imines.

Results and Discussion

The reaction of imines with cyclic anhydrides has been used for the synthesis of pyrrolidines, piperidines and isoquinolines (12) (Figure 1). The proposed mechanism involves initial iminolysis of the anhydride followed by ring closure. The resulting isoquinoline contains a 4-carboxy substituent which can be further derivatized. Isoquinoline formation has been demonstrated with a variety of imines formed from primary amines ranging from methyl to t-butyl and anilinyl. The use of substituted benzaldehydes indicated that the reaction proceeded with both electron-rich (p-dimethylamino substituent) and electron-deficient (p-nitro substituent) imines. The generality of the reaction indicated in the literature made it an attractive candidate for combinatorial library synthesis.

[1]Corresponding authors

The stereochemistry of the substituents on the ring formed in the reaction vary widely. This is likely due to the steric and electronic properties of the intermediate imine. However, isomerization to the more stable trans configuration has been demonstrated using either acid or base catalysis. This conversion to a known stereochemistry was deemed desirable due to the extreme range of cis/trans products observed in test reactions ranging from 1:9 to 9:1. Solution tests of isoquinoline formation were successfully performed with imines formed from glycine ethyl ester and benzaldehyde. Analysis of the coupling constants of H-3 and H-4 indicated that the product was a 9:1 mixture of cis and trans diastereomers. Treatment with 1 N NaOH (aqueous) resulted in conversion to >95% trans isomer. Treatment with 75:20:5 TFA/dichloromethane/water for 135 minutes resulted in neither isomerization nor decomposition, indicating that resin cleavage could be performed safely using the cleavage conditions required for the RINK linker (13).

The feasibility of solid-phase syntheses were initially carried out using Tentagel - NH_2 resin carrying a RINK linker. Good yields of isoquinoline were obtained from the resin-bound imine derived from glycine and benzaldehyde. Imine formation was driven by use of the dehydrating agent trimethyl orthoformate (14). Reactions performed with substituted benzaldehydes confirmed that electron-rich and electron-deficient imines resulted in clean conversion to isoquinolines as did ortho-substituted benzaldehydes. The use of amines bearing a 1-substituent, however, resulted in competitive formation of the homophthalic amide of these amines. This side reaction appears to be related to the substituent's steric bulk. Bearing this constraint in mind, a small array synthesis was performed on solid phase in order to test the generality of the isoquinoline formation.

A 4 x 6 array of isoquinolines was synthesized as shown in Figure 2. The reaction of a single aldehyde with a mixture of amino acids on resin was also tested. After cleavage from the resin, the isoquinolines were isomerized to the trans product with 1 N NaOH and then back-extracted into organic solvent after acidification of the aqueous layer. Reasonable yields (>50%) of all expected isoquinolines were obtained except for those formed from 3-hydroxybenzaldehyde. The low recovery of these isoquinolines was due to substantial solubility of the product in acidic water and could be increased by addition of NaCl to the water layer and repeated organic extractions as seen in the yield for the mixed amino acids reacted with this aldehyde. NMR indicated >95% trans configuration of the ring substituents after the base treatment. Mass spectrometry on the pool containing mixed amino acids showed the presence of all expected isoquinolines.

Synthesis of the isoquinoline array identified three areas requiring further study: an improved isomerization method to avoid post-cleavage extraction, a method for carboxyl derivatization to provide a third site of diversity, and a simpler cleavage procedure more suited to multiple sample handling. The latter question was readily solved as the isoquinoline synthesis on MBHA resin was shown to be identical to that on Tentagel and no decomposition was observed after HF treatment, allowing the use of a multiple sample HF cleavage apparatus (15). On-resin activation of the carboxyl was achieved by treatment with O-(7-azabenzotriazol-1-yl)-1,1,3,3-tetramethyluronium hexafluorophosphate (HATU) (16) and high yield conversion to primary and secondary amides as well as hydrazides was demonstrated using a double activation/coupling procedure (Figure 3). Ester formation, however, proceeded in low yield with alcohol nucleophiles and amide formation was minimal with the poorly nucleophilic p-nitroaniline. A unexpected feature of this coupling protocol was the discovery that treatment with 1 M amines or hydrazines concomitantly achieved the desired isomerization to the trans diastereomer. Synthesis of a combinatorial library was now possible with the general conditions for isoquinoline formation and derivatization having been established. Library synthesis was performed on MBHA resin partitioned in porous polypropylene packets for ease of handling (17).

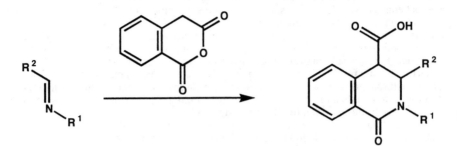

Figure 1. Synthesis of tetrahydroisoquinolines from imines.

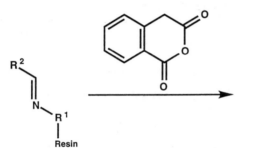
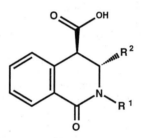

	3-Amino-propionic Acid	4-Amino-butyric Acid	6-Amino-hexanoic Acid	Glycine	Mixed Amino Acids
Benzaldehyde	89%	97	132	68	61
4-Methoxy-benzaldehyde	87	58	78	51	93
3,5-Dimethoxy-benzaldehyde	75	83	82	47	84
4-Cyano-benzaldehyde	50	69	74	52	89
2-Bromo-benzaldehyde	82	84	83	104	94
3-Hydroxy-benzaldehyde	11	11	15	18	112

Figure 2. Solid-phase array synthesis of isoquinolines.

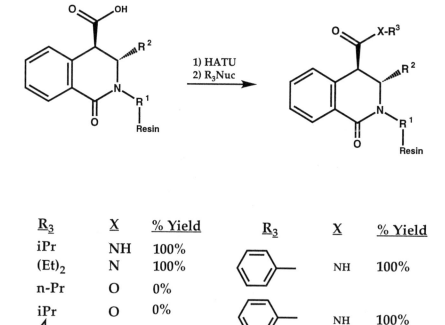

R_3	X	% Yield
iPr	NH	100%
(Et)$_2$	N	100%
n-Pr	O	0%
iPr	O	0%
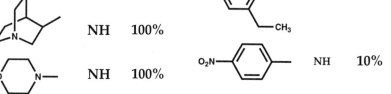	NH	100%
	NH	100%

R_3	X	% Yield
	NH	100%
	NH	100%
	NH	10%

Figure 3. On-resin carboxyl activation and coupling.

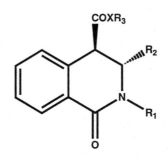

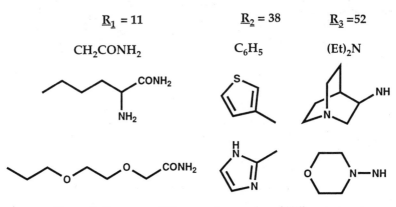

Figure 4. Summary of library and examples of "R" groups.

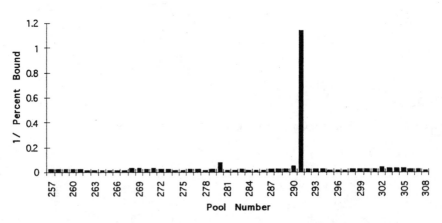

Figure 5. Sigma and opioid receptor binding assays of library pools.

The isoquinoline library was prepared using 11 amino acid building blocks, 38 aldehydes, and 51 amine nucleophiles. Library preparation used the divide, coupled and recombine method (16) whereby 11 amino acids were separately coupled to MBHA resin, the resins were mixed and then split into 38 equal size pools. Each mixed amino acid pool underwent imine formation with a single aldehyde and was then reacted with homophthalic anhydride. The resulting 38 isoquinoline pools were then again mixed and divided into equal portions for coupling to amine nucleophiles. The uncoupled 4-carboxy substituent was included as the 52nd substituent at position 3. The overall number of compounds needs to account for the fact that the all trans products are a 1:1 ratio of enantiomeric isoquinolines (diastereomeric products in compounds where any of the building blocks also possessed chirality). This combination therefore provided a library of 11 x 38 x 52 x 2 = 43,472 compounds (Figure 4). This library consisted of pools of 836 compounds. The library specifically consisted of 47 pools of 836 and 5 pools of 1672 compounds; the larger pools were due to the use of a racemic amine in the last step generating two sets of diastereomers.

We believe that the preferable method for screening combinatorial libraries is as soluble mixtures (1-4, 6, 15, 17-19, 21); this allows the investigation of large numbers of compounds while avoiding the synthetic limitations of tagged libraries and the drawbacks of assaying solid-phase bound libraries. The screening of combinatorial library compound mixtures can be compared to the assay of natural product extracts, a well established method for the identification of therapeutically active lead compounds. In this comparison the synthetically generated mixture has two distinct advantages over the natural product extract: the compounds are present in approximately equimolar ratios and the identification, synthesis and optimization of active components is accelerated due to knowledge of the mixture's contents and means of synthesis .

Production of pools containing more than 50-100 compounds currently precludes analytical confirmation of the presence of every expected product although new techniques and instrumentation may soon extend this ability to much larger pools. We employ several approaches to maximize and measure the fidelity of our libraries. Optimization of reaction conditions is designed to cover as wide a range of sterically and electronically demanding reactants as possible. Reaction vessels for library synthesis contain both a resin packet of the mixed resin for the actual library and a packet containing a single resin-bound precursor which is later analyzed to ensure that the reaction occurred as planned. Finally, while it may be impossible to identify every compound in the mass spectrum of a large mixture, the modification of that mixture can often be confirmed by observation of the mass shift of the overall molecular weight distribution. For example, if a large pool of isoquinoline carboxylic acids is activated and coupled to an amine of molecular weight 118 the overall mass pattern should shift by 100 mass units (118 minus 18 for the loss of water in the condensation reaction).

Examination of the control bags for the 38 aldehyde reactions revealed that all reacted with the resin-bound amine and homophthalic anhydride to form isoquinolines. Mass spectrometry indicated that one of the aldehydes, 5-(hydroxymethyl)furfural, produced in good yield a substance with molecular weight 90 mass units greater than the expected isoquinoline. NMR spectroscopy identified this compound as the Friedel-Crafts alkylation product of the expected isoquinoline and the anisole scavenger used in the HF cleavage. All of the amine nucleophiles gave good conversion of a control isoquinoline carboxylic acid to the expected amides or hydrazides; mass spectra of the actual library pools showed the expected shift in mass of the molecular weight distribution. The library pools were obtained with average and median yields of 69%.

As has been described previously (18-20), we tested the isoquinoline library in kappa and mu opioid receptor binding assays and in a sigma receptor binding assay. Differing specificity of the library was observed for the three receptors (Figure 5). The greatest activity was observed in the sigma receptor binding assay for pool 291

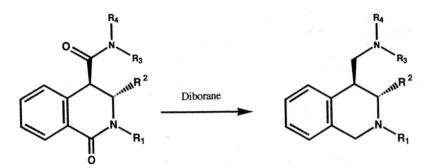

Figure 6. Libraries from libraries: reduction to 4-aminomethylisoquinolines

containing adamantanemethylamine amides of the mixed isoquinolines. Deconvolution of pools active in these assays is in progress.

The generation of "libraries from libraries" (3) is an efficient method of generating increased diversity. The "libraries from libraries" concept involves the blanket transformation of one library into a related library exhibiting significantly different physical properties. One example of this is the transformation of a tetrapeptide library into a polyamine library via reduction of the peptide's amide bonds (3). The feasibility of diborane reduction of the 4-carboxamidotetrahydroisoquinoline library to provide a 4-aminomethyltetrahydroisoquinoline library was examined with test compounds (Figure 6). Complete reduction of the isoquinolines was achieved.

The successful synthesis of this library and observation of bioactivity have led us to plan a significantly larger tetrahydroisoquinoline library. This library will also be reduced to form a second library which will exhibit significantly different biological properties due to its modified ring conformation and positive charge at physiological pH. We are also preparing a version of the tetrahydroisoquinoline library in positional scan format to utilize this powerful deconvolution method (21).

Acknowledgments. We would like to thank Phibun Ny, Thomas Hayes, and Yazhong Pei for assistance and helpful discussions. This work was funded by Houghten Pharmaceuticals, Inc. San Diego, CA.

References

1. Blondelle, S. E.; Perez-Paya, E.; Dooley, C. T.; Pinilla, C.; Houghten, R. A. *Trends Anal. Chem.* **1995**, 14, 83-92.
2. Houghten, R. A.; Pinilla, C.; Blondelle, S. E.; Appel; J. A.; Dooley, C. T.; Cuervo, J. H. *Nature* **1991**, *354*, 84-6.
3. Ostresh, J. M.; Husar, G. M.; Blondelle, S. E.; Doerner, B.; Weber, P. A.; Houghten, R. A. *Proc. Natl. Acad. Sci. USA* **1994**, *91*, 11138-42.
4. Houghten, R. A.; Appel, J. R.; Blondelle, S. E.; Cuervo, J. H.; Dooley, C. T.; Pinilla, C. *Biotechniques*, **1992**, *13*, 412-21.
5. Zuckermann, R. N.; Martin, E. J.; Spellmeyer, D. C.; Stauber, G. B.; Shoemaker, K. R.; Kerr, J. M.; Figliozzi, G. M.; Goff, D. A.; Siani, M. A.; Simon, R. J.; Banville, S. C.; Brown, E. G.; Wang, L.; Richter, L. S.; Moos, W. H. *J. Med. Chem.* **1994**, *37*, 2678-2685.
6. Cuervo, J. H.; Weitl, F.; Ostresh, J. M.; Hamashin, V. T.; Hannah, A. L.; Houghten, R. A. *Polyalkylamine Chemical Combinatorial Libraries*; Maia, H. L. S., Ed.; ESCOM Science Publishers: Braga, Portugal, 1994, pp 465-466.
7. Chen, C.; Randall, L. A. A.; Miller, R. B.; Jones, A. D.; Kurth, M. J. *J. Am. Chem. Soc.* **1994**, *116*, 2661-2662.

8. DeWitt, S. H.; Kiely, J. S.; Stankovic, C. J.; Schroeder, M. C.; Cody, D. M. R.; Pavia, M. R. *Proc. Natl. Acad. Sci. USA* **1993**, *90*, 6909-6913.
9. Bunin, B. A.; Ellman, J. A. *J. Am. Chem. Soc.* **1992**, *114*, 10997-10998.
10. Holmes, C. P.; Chinn, J. P.; Look, G. C.; Gordon, E. M.; Gallop, M. A. *J. Org. Chem.* **1995**, *60*, 7328-7333.
11. Goff, D. A.; Zuckermann, R. N. *J. Org. Chem.* **1995**, *60*, 5748-5749.
12. Cushman, M.; Madaj, E. J. *J. Org. Chem.* **1986**, *52*, 907-915.
13. Rink, H. *Tetrahedron Lett.* **1987**, *28*, 3787.
14. Look, G. C.; Murphy, M. M.; Campbell, D. A.; Gallop, M. A. *Tetrahedron Lett.* **1995**, *36*, 2937-2940.
15. Houghten, R. A.; Bray, M. K.; DeGraw, S. T.; Kirby, C. J. *Int. J. Pep. Pro, Res.*, **1986**, *27*, 673-78.
16. Carpino, L. A.; El-Faham, A.; Minor, C. A.; Albericio, F. *J. Chem. Soc., Chem. Commun.* **1994**, 201-203.
17. Houghten, R. *Proc. Natl. Acad. Sci., USA* **1985**, *82*, 5131-5135.
18. Dooley, C. T. ; Houghten, R. A. *Life Sci.* **1993**, *52*, 1509-17.
19. Dooley, C. T.; Houghten, R. A. *Analgesia*, **1995**, *1*, 2019-20.
20. de Costa, B. R.; Bowen, W. D.; Hellewell, S. B.; Walker, J. M.; Thurkauf, A.; Jacobson, A. E.; Rice, K. R. *FEBS Lett.* **1989**, *251*, 53-58.
21. Pinilla, C.; Appel, J. R.; Blondelle, S. E.; Dooley, C. T.; Eichler, J.; Ostresh, J. M.; Houghten, R. A. *Drug Dev. Res.* **1994**, *33*, 133-145.

Chapter 7

Solid-Phase and Combinatorial Synthesis of Heterocyclic Scaffolds

Dihydropyridines, Pyridines, and Pyrido[2,3-*d*]pyrimidines

Dinesh V. Patel, Mikhail F. Gordeev, Bruce P. England, and Eric M. Gordon

Affymax Research Institute, 3410 Central Expressway, Santa Clara, CA 95051

A: Introduction

Combinatorial chemistry involves generation of molecular diversity through covalent combination of members of a set of structurally varied chemical building blocks.[1] This has been applied in fundamentally two different ways. The first and simpler approach involves synthesis of linear, homo-oligomeric libraries through iteration of a single coupling chemistry employing a single set of building blocks, e.g. repeatative amide coupling reactions with a set of amino acid building blocks to synthesize peptide libraries. While such peptidic libraries certainly helped to validate the concept of molecular diversity, peptidic molecules are expected to have limited scope in the field of drug discovery because of their poor bioavailability. Thus, more recently, attention has been diverted to the alternative strategy of combinatorial synthesis of small molecule, cyclic, non-polymeric scaffolds utlizing several different sets of building blocks and employing multiple connection chemistries.[2] The latter approach leads to generation of a diverse collection of small sized, non-peptidic, heterocyclic molecules, the type of which have historically found to be most successful as drugs.

Herein, we describe new synthetic methodologies for solid phase combinatorial synthesis of dihydropyridines (DHPs), pyridines and pyrido-[2,3-d]-pyrimidines. Structures or substructures of this class of nitrogen heterocycles have emerged in a wide variety of bioactive molecules and therapeutic agents.

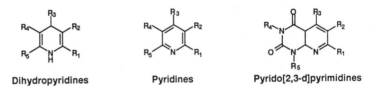

Dihydropyridines   Pyridines   Pyrido[2,3-d]pyrimidines

B: Dihydropyridines (DHPs)

The dihydropyridine core nucleus **1** has been found to be present in numerous molecules spanning a wide range of bioactivity, the most successful examples being the calcium channel blocker antihypertensive drugs such as Nifedipine, Nitrendipine, and Nimodipine.[3] A synthetic route was envisioned wherein the point of attachment for the heterocyclic DHP moiety **1** to the solid support is through its nitrogen atom.

1054–7487/96/0058$15.00/0

Retrosynthetic disconnections identify enamino esters **2** and arylidene 1,3-dicarbonyls **3** as the penultimate fragments, which in turn can be assembled from *beta*-ketoesters **4**, *beta*-dicarbonyls **5**, aromatic aldehydes **6**, and the amine resins. While a large variety of these building blocks (BBs) are commercially available, major uncertainties that we faced at the onset were the feasibility of preparing immobilized enamino esters **2**, effectiveness of Hantzsch type heterocyclization to form the core nucleus, and the stability of oxidatively labile DHP ring **1** on solid support to final cleavage conditions. In order to address these issues, solid phase synthesis of Nifedipine was undertaken as a model study.

Solid Phase Synthesis of Dihydropyridines (DHPs): Retro-Synthetic Analysis

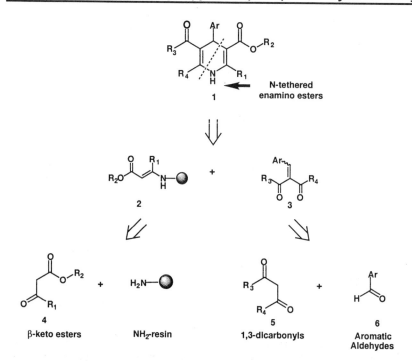

1) Solid Phase Synthesis of Nifedipine: A Model Study

Treatment of acid cleavable H_2N-PAL or H_2N-Rink resin in CH_2Cl_2 with an excess of ethyl acetoacetate in the presence of catalytic amounts of pTsOH and 4A° molecular seives as dehydrating agent for 48 h at room temperature formed the enamino ester **7**. Typically, the reaction was allowed to proceed until Kaiser test was practically negative, indicating total disappearance of free amine from the resin. Next, the resin bound enamino ester **7** was treated with preformed arylidene ester **8** and subjected to standard Hantzsch type heterocyclization conditions. The reaction at this stage was expected to form the Michael adduct **9** and undergo a N-C cyclization to form Nifedipine **10**. Instead, the cleaved product, which had the same molecular weight by MS as Nifedipine, was identified by ^1H NMR to be the carbocyclic cyclohexadiene intermediate **11**, apparently arising from an aldol type C-C cyclization. This result underscores the importance of thorough characterization in such studies aimed at developing new chemistries on solid support. Probing the reaction mechanism led us to speculate that the use of a base as an additive or as a solvent may help isomerize the imine **9** to the conjugated enamime, and that the latter may then be more prone to the desired N-C cyclization. After some experimentation, the desired product Nifedipine **10** was indeed obtained at room temperature by using pyridine as a solvent and employing 4A° molecular seives as the dehydrating agent (70% yield, 80% purity).

Solid Phase Synthesis of Nifedipine: N-Tethered Enamino Ester Route

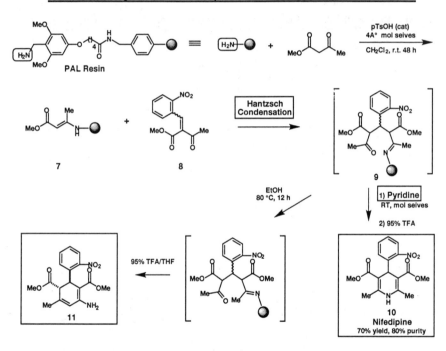

Having successfully accomplished the synthesis of Nifedipine **10** on solid support, an extensive chemistry rehearsal study was undertaken to optimize the reaction conditions with respect to parameters such as resin, solvent, temperature, cyclization and cleavage conditions, and the compatibility of electronically and structurally diverse building blocks to the solid phase synthesis protocol. The key results are summarized below.

1) Resin:
Although PAL resin was employed initially, the more acid labile Rink resin (cleavable by 3% TFA) was prefered in subsequent studies.
2) Cyclization:
As discussed above, pyridine is the most effective solvent for desired Hantzsch type heterocyclizations. Two different methods of DHP synthesis were developed.
a) Three component cyclization: Immobilized enamino ester **2** is treated with a mixture of aromatic aldehyde **6** and 1,3-dicarbonyl building blocks **5**. This route is more convenient and relies on *in situ* generation of arylidene dicarbonyls **3**, and works for electron poor or neutral aldehydes.

b) Two component cyclization: It involves reaction of enamino ester **2** on solid support with arylidene dicarbonyls **3** preformed separately by treatment of aromatic aldehydes with 1,3-dicarbonyls. This method offers better scope for diversity and works for both electron rich and poor aromatic aldehydes.

2- and 3-Component Methods for Synthesis of Dihydropyridines (DHPs)

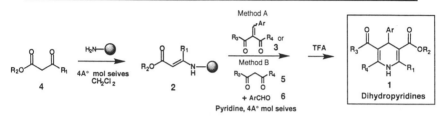

3) Diversity: Both electron rich and poor aromatic aldehydes can be accomodated by appropriate choice of cyclization method. In general, there is good tolerance for linear alkoxycarbonyl substituents at positions 3 and 5 (R_2 and R_3), and linear substituents are acceptable at positions 2 and 6 (R_1 and R_4).

2) Model Study: Preparation of a "9 member" DHP Library

Having worked out the scope & limitations of solid phase DHP synthesis, we next commenced on rehearsing the library synthesis. Treatment of Rink-NH_2 resin with methyl, allyl and benzyl acetoacetate leads to 3 different enamino esters which are pooled and then split into 3 portions. Each pool is treated with a different arylidene keto ester to give a 9 member library in the form of 3 pools (A, B, and C) with 3 members/pool. As expected, HPLC analysis indicated 3 major components per pool (data not shown). With respect to pool A, authentic samples of expected components were individually synthesized and used for co-injection study to confirm the identity of the 3 components (data not shown). MS analysis also gave the molecular ion peaks for desired mixture of compounds for each pool. This library rehearsal exercise assured us of the reliability and reproducibility of the DHP chemistry in the 'split pool' format, and set the stage for preparation of larger sized libraries.

3) Construction of a "100 member" DHP Library

Preparation of a larger size DHP library was undertaken to screen for calcium blockade activity as a test study, and for eventually screening against a range of enzyme and receptor targets in a high throughput screening format. In the first step, 10 different *beta*-keto esters representing an adequate degree of structural diversity were employed, and the resulting enamines were pooled together. In the second step, a 3-component heterocyclization was employed. Thus, methyl acetoacetate was chosen as the 1,3-dicarbonyl component and held constant, and diversity was derived from choice of 10 different aromatic aldehydes. The end result was a 100 member DHP library derived from employing 10 *beta*-keto esters in the first step and 10 aromatic aldehydes in the second step. The library is in the form of 10 different pools, with each pool bearing a distinct, single aryl group and 10 different R_1/R_2 substituents derived from the 10 *beta*-keto esters employed for forming enamino esters.

Construction of a "9 member" Model DHP Library

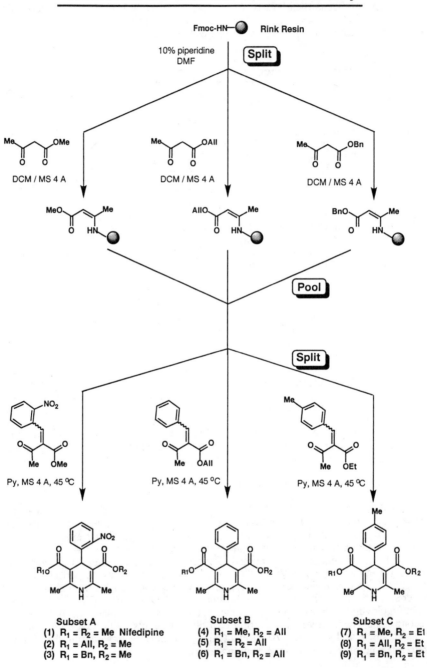

Subset A
(1) $R_1 = R_2 = Me$ Nifedipine
(2) $R_1 = All, R_2 = Me$
(3) $R_1 = Bn, R_2 = Me$

Subset B
(4) $R_1 = Me, R_2 = All$
(5) $R_1 = R_2 = All$
(6) $R_1 = Bn, R_2 = All$

Subset C
(7) $R_1 = Me, R_2 = Et$
(8) $R_1 = All, R_2 = Et$
(9) $R_1 = Bn, R_2 = Et$

Design and Construction of a 100 member DHP Library

Step I: Preparation of Immobilized Enamino Esters

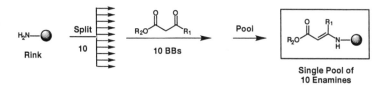

β-Keto Ester Building Blocks

Step II: A 3 component Hantzsch Cyclocondensation

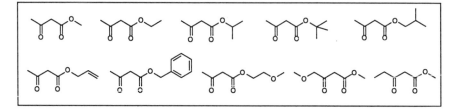

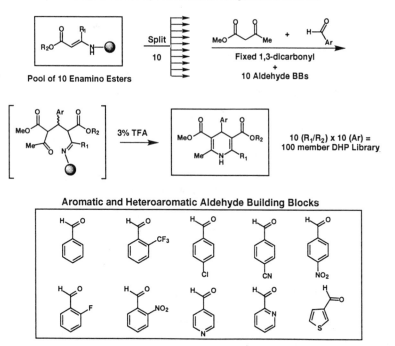

Aromatic and Heteroaromatic Aldehyde Building Blocks

4) Screening of the "100 member" DHP Library

Screening of the 10 pools for calcium blockade activity using a cortex membrane binding assay[4] identified pools **AF-0591-7** and **AF-0591-3** derived from *ortho*-nitrobenzaldehyde and *ortho*-fluorobenzaldehyde repsectively as the most active pools. For deconvolution, parallel synthesis of individual members of a pool was performed and rough I_{50} estimanations were made using the crude, unpurified samples. The compounds of interest were then purified by HPLC, fully characterized, and used for precise I_{50} value determinations. Thus, deconvolution of the *ortho*-nitrobenzaldehyde pool identified Nifedipine (**AF-0591-7-1**, I_{50} = 18 nM) and the closely related ethyl ester analog (**AF-0591-7-2**, I_{50} = 12 nM) as some of the most active members.

Case Study: Screening of 100 member DHP Library to Identify Calcium Channel Blockers

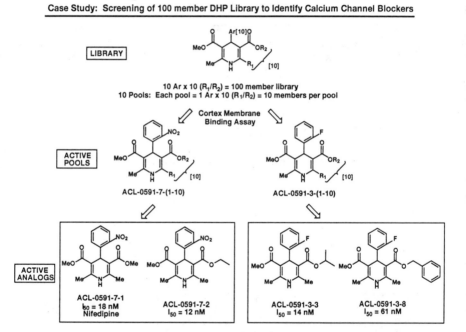

A better spread in activity was realized with the *ortho*-fluorobenzaldehyde pool as evidenced by the activity of isopropyl acetate analog (**AF-0591-3-3**, I_{50} = 14 nM) vs. the benzyl ester (**AF-0591-3-8**, I_{50} = 61 nM) analog. In summary, we were successfully able to identify a known, active drug entity such as Nifedipine and other potent analogs out of a 100 member library. This study reassures us of the adequate quality of DHP analogs in the library prepared via this synthetic route, and helps us validate the concept of combinatorial chemistry aided drug discovery.

C: Solid Phase Synthesis of Pyridines

An attractive site for further synthetic manipulations of the DHP nucleus is its secondary nitrogen group, which for example can be oxidized to pyridines or subjected to other transformations such as alkylations and acylations. Our present N-tethered synthetic route to DHPs would however preclude us from extending the combinatorial scope and utility of DHP nucleus in this manner. This prompted us to explore an alternate synthesis of such heterocyclic molecues, wherein the attachment on solid support is via the ester side chain. An important difference from the previous route is that the nitrogen atom of the DHP ring system formed on solid support will now be available for oxidation to pyridines and other synthetic modifications. Thus, this alternative strategy commenced with acetoacetylation which was conveniently performed by treatment of alcohol Tentagel resin with diketene (cat. DMAP, CH_2Cl_2, -50 °C to rt). The loading of keto ester **12** on resin can be evaluated by cleavage and cyclization by treatment with ethanolic hydrazine and spectrophotometric quantitative estimation of the resulting pyrazolinone **13**. Employing ^{13}C enriched building blocks, the course of reaction on Wang, Sasrin or Tentagel resin could be monitored by gel phase ^{13}C NMR. Knoevenagal type condensation of immobilized keto ester **12** with ^{13}C labelled benzaldehyde forms arylidene ester **14**, as witnessed by a clean signal for vinylic carbon at 139 ppm. Treatment of **14** with enamino ester **15** under typical Hantzsch cyclization conditions (ethanol, reflux) led to the assembly of DHP nucleus **16** on solid support, as confirmed by a complete shift of the ^{13}C signal from vinylic (139 ppm) to allylic (39 ppm) region. Oxidation of the DHP ring to pyridine proceeds cleanly on solid support upon treatment with ceric ammonium nitrate (CAN), and the ^{13}C signal shifts downfield as expected into the aromatic region (145 ppm). Finally, cleavage with TFA affords pyridine-3-carboxylic acids **17** in good yield and purity.

SPS of DHPs and Pyridines from Immobilized β-Keto Esters

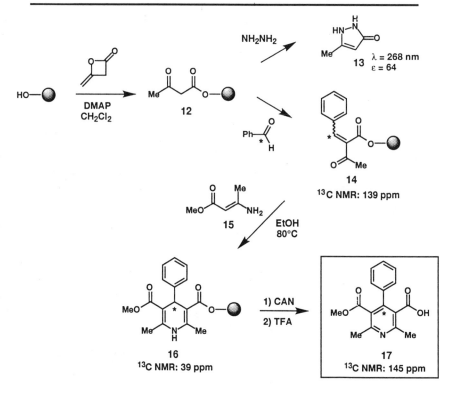

Gel Phase ^{13}C NMR Study of SPS of Dihydropyridines and Pyridines

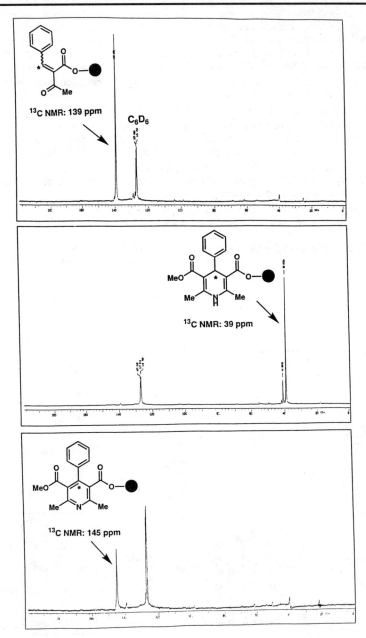

The procedure is fairly general and has been successfully applied to preparation of a wide variety of nicotinic acid derivatives **21**. Unlike the previous route, the choice of aldehyde **18** building blocks is not restricted to aromatic aldehydes, and aliphatic aldehydes have also been successfully employed (R_1, see table, entry 9 and 10). Additionally, the 1,3-enaminoketone component also accomodates incorporation of enamines **20** derived from symmetrical 1,3-diketones (R_2 and R_3, see table, entry 13 and 14) besides the *beta*-keto esters.

Rehearsing the Chemistry for Solid Phase Synthesis of Nicotinic Acids

Entry #	R_1	R_2	R_3	HPLC product purity,[a] %	MS
1	Ph(C^{13})	MeO	Me	80	O'K
2	Ph(C^{13})	^{i}PrO	Me	95	O'K
3	p-HOOCPh	MeO	Me	99	O'K
4	o-FPh	MeO	Me	98	O'K
5	2-naphthyl	MeO	Me	98	O'K
6	4-Py	^{i}PrO	Me	81	O'K
7	m-O$_2$NPh	MeO	Me	95	O'K
8	p-MeOPh	^{i}PrO	Me	90	O'K
9[b]	n-Hex	MeO	Me	90	O'K
10[b]	S~	MeO	Me	70	O'K[c]
11[b]	Ph	MeO	Me	90	O'K
12[b]	N~~OPh-p	MeO	Me	70	O'K[d]
13[b]	Ph	Me	Me	90	O'K
14[b]	Ph	CH$_2$CMe$_2$CH$_2$		90	O'K

Notes: [a]Detection at 220 nm. [b]Prepared using synthsizer. [d]Dealkylation product's ion $[M - R_1 + 2H]^+$ was also detected.

D   Solid Phase Synthesis of pyrido[2,3-d]pyrimidines:

The development of DHP and pyridine chemistry provides numerous opportunities for further chemical modification of intermediates and final scaffolds to expand the portfolio of heterocyclic libraries. Once the fundamental chemistry is worked out, the structural complexity of scaffolds is not a limiting factor for creating heterocyclic combinatorial diversity. As an example, replacing the enamino-ketone component with 6-aminouracil **22** in the immobilized *beta*-keto ester route led to the formation of pyrido[2,3-d]pyrimidine **23** in excellent purity. Interestingly, it has been reported that syntheses of this heterocycle using unsubstitued aminouracil could not be effected in solution because of the lack of reactivity of enamine.[5] Presumably, the solid phase route is benefiting from the principle of excess reagents helping to drive it to completion in good yield and purity.

Solid Phase Synthesis of Pyrido[2,3-d]pyrimidines

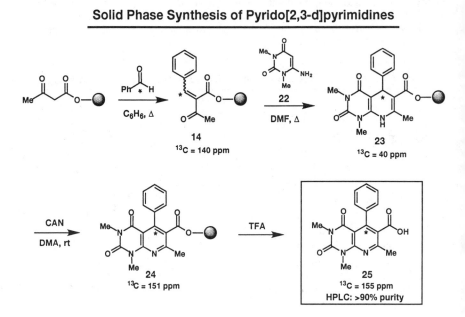

E   Conclusion:

The strong precedence of nitrogen heterocycles as drugs has prompted a desire for development of solid phase synthetic methodologies for such class of molecules. This in turn can be expected to facilitate the construction of corresponding libraries and enable one to evaluate thousands of analogs in a combinatorial chemistry based high throughput screening format. The dihydropyridine nucleus was chosen as a starting point in this study. The solid phase synthesis method development was followed by preparation of libraries, and their screening for calcium channel blockade activity led to the identification of expected as well as new potent ligands. An alternate route to this nucleus was also enabled which allowed easy construction of corresponding pyridine libraries thorugh an oxidation step. The synthetic events on solid support can be conveniently monitored through the use of ^{13}C gel phase NMR. Through the choice of appropriate building blocks, new scaffolds can be generated, e.g. uracils instead of enamino ketones will form bicyclic nitrogen heterocycles such as pyridopyrimidines. Similarly, various intermediates and scaffolds can serve to provide further points of diversity. In the near future, we can anticipate the integrated development of new connection chemistries, better instrumentation and high throughput assay systems to play a dominant role in the new era of drug discovery.

Acknowledgements:

We would like to thank Ms. Jie Wu and Ms. Supriya Jonnalagadda for their technical help in this research project.

References:

1. (a) Gallop, M. A.; Barrett, R. W.; Dower, W. J.; Fodor, S. P. A.; Gordon, E. M. *J. Med. Chem.* **1994**, *37*, 1233.   (b) Gordon, E. M.; Barrett, R. W.; Dower, W. J.; Fodor, S. P. A.; Gallop, M. A. *J. Med. Chem.* **1994**, *37*, 1385.   (c) Gordon, E. M. *Current Opinion in Biotechnology.* **1995**, *6*, 624-631.   (d) Gordon, E. M.; Gallop, M. A. *Acc. Chem. Res.* **1996**, *29*, 144-154.   (e) Patel, D. V.; Gordon, E. M. *Drug Discovery Today.* **1996**, 1, 134-144.

2. Some recent examples are listed below: (a) Kick, E. K.; Ellman, J. A. *J. Med Chem.* **1995**, *38*, 1427.   (b) Holmes, C. P.; Jones, D. G. *J. Org Chem.* **1995**, *60*, 2318.   (c) Campbell, D. A.; Bermak, J. C.; Burkoth, T. S.; Patel, D. V. *J. Am. Chem. Soc.* **1995**, *117*, 5381.   (d) Murphy, M. M.; Schullek, J. R.; Gordon, E. M.; Gallop, M. A. *J. Am. Chem. Soc.* **1995**, *117*, 5381.   (e) Ruhland, B.; Bhandari, A.; Gordon, E. M.; Gallop, M. A. *J. Am. Chem. Soc.* **1996**, *118*, 253.   (f) Gordeev, M. F.; Patel, D. V.; Gordon, E. M. *J. Org. Chem.* **1996**, *61*, 924.

3. Tentagel is a cross linked polystyrene resin grafted with polyethylene glycol (PEG). The PEG linker is symmetrical in nature and results in a single sharp resonance at 70 ppm, whereas the polystyrene matrix is essentially NMR invisible, thereby providing a convenient handle to monitor reactions on solid phase by using ^{13}C enriched samples Look, G. C.; Holmes, C. P.; Chinn, J. P.; Gallop, M. A. *J. Org. Chem.* **1994**, *59*, 7588.

4. Boecker, R. H.; Guengerich, F. P. *J. Med. Chem.* **1986**, *29*, 1596.

5. Kajino, M.; Meguro, K. *Heterocycles.* **1990**, *31*, 2153.

Chapter 8

Polymer-Supported Chemistry: Synthesis of Small-Ring Heterocycles

A. M. M. Mjalli and B. E. Toyonaga

Ontogen Corporation, 2325 Camino Vida Roble, Carlsbad, CA 92009

The production of libraries of small organic molecules is a core technology at Ontogen. The methods by which we synthesize and assay individual compounds within the libraries, as well as an overview of some of the chemistry involved has been previously described. To continue the description, recently developed polymer-supported chemistries will be discussed. For example, the utility of a pyrrole synthesis in which four substituents may be independently manipulated to produce vast libraries (greater than 10^8) will be illustrated. In addition, chemistries used to synthesize libraries of other small-ring heterocycles, such as imidazoles, will be reviewed. Finally, initial biological assay data from current therapeutic discovery projects will be presented.

High speed combinatorial synthesis of nonpeptidyl, small molecule chemical libraries is emerging as a powerful tool in the discovery of biologically active molecules (1). Most recently, several deconvolution strategies of combinatorial libraries using radio frequency encoded (1d-2) and spatially dispersed (3) combinatorial libraries have been described. The necessary tools for compound library synthesis, including hardware and software, are important features to facilitate high speed synthesis (2a). Since libraries of large numbers of compounds can be synthesized, diversity within such libraries is highly desirable and constitutes a pivotal aspect of drug discovery.

Diversity in Theory

The availability of tools (chemistry software and computer hardware) allow chemists to generate "virtual" structures through combinatorial chemical methods using readily available starting materials. Typically, the number of computer-generated compounds is larger than the number of compounds likely to be synthesized by several orders of magnitude. Use of commercially available software and published methods (*4*) can eliminate reagent "inputs" with undesirable chemical properties such as covalently bonded metals, alkylating agents, isotopomers, low stability compounds and Michael addition reagents. Reagents of high molecular weight (>600) could also be eliminated. Typically, methods used to quantify molecular diversity are based on the following molecular characteristics:

- lipophilicity
- size
- shape and branching
- chemical functionality
- specific binding features

Consideration of these factors would lead to the actual synthesis of a subset of the virtual library, a smaller representative library. As such, this library might provide an acceptable representation of chemical diversity.

Diversity in Practice

In contrast, the most complete approach would involve the synthesis of all possible structures. Two essential tools are necessary in this case:

- Technology: The necessary hardware and software (*2a*) with which a wide range of organic chemical reactions could be carried out. These tools should allow chemists to conduct reactions based on multicomponent condensation reactions and/or sequential addition reactions under different reaction conditions such as high/low temperature and inert atmosphere conditions.
- Synthetic Methodology: The library size and its diversity is dependent upon the synthetic strategy. A highly efficient method, which utilizes a wide range of readily available chemical reactant "inputs" is necessary.

Overall, for the application of combinatorial chemistry to SAR based lead compound optimization, milligram quantities of pure compound should be prepared (*2a*). At Ontogen, compounds are synthesized on a solid support and cleaved in the final step to produce SDCLs. In this way, relatively pure compounds are prepared without the need for exhaustive chromatographic purification procedures. All of the methods developed are amenable to automated synthesis. These systems, collectively known as the OntoBLOCK system, have been described elsewhere (Cargill, J.F.; Maiefski, R. R.; Toyonaga, B. E. *Laboratory Robotics and Automation* , **1996**, in press.).

Herein, the generation of chemical libraries using methods based on four component condensation reactions and post four component condensation reaction transformations are described.

Four Component Condensation Reactions

The four component condensation reaction between an isocyanide (R_1NC), an acid (R_3COOH), an amine(R_2NH_2) and an aldehyde (R_4CHO) provides N-alkyl-N-acyl-α-aminoamide derivatives (5) (Figure 1). The reaction involves imine formation $R_2HN=CR_4$ followed by addition of the isocyanide R_1NC and the acid R_3COOH to give the intermediate 1 which undergoes acyl transfer to provide 2. It has become clear that almost any combination of the possible four components may be used in this reaction. Sterically hindered components can also be used. Electronically and sterically diverse acid components, 1°, 2°, aliphatic/aromatic amines, hydrazines, hydroxyl amines and their derivatives can be used in this 4-component reaction (5). Notwithstanding the limited number of readily available isocyanides (twelve), millions of highly complex compounds of general formula 2 may be generated based on the hundreds of commercially available amines, carboxylic acids and aldehydes. Our strategy is (i) to take advantage of this one step reaction to synthesize large numbers of highly diverse compounds with significant variations of R_2, R_3 and R_4 and (ii) to convert these dipeptide derivatives 2 to pharmaceutically acceptable compounds such as imidazoles (6), pyrroles (7) and oxazoles. By this strategy, directed diversity is realized in the small heterocycles in a manner which would be difficult, and sometimes impossible, by any other approach.

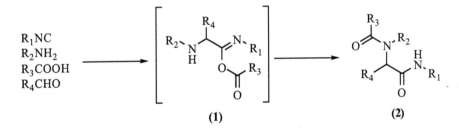

Figure 1. Four Component Condensation Reaction

Synthetic Strategies and Goals. The strategy of our synthesis involves the following:

- the synthesis of N-alkyl-N-acyl-α-amino amides 2 on solid support;
- the development of a convertible isocyanide in the four component condensation reaction and
- the conversion of N-alkyl-N-acyl-α-amino amides derivatives 2 or 5 to highly substituted pyrroles, imidazoles etc. as outlined in Figure 2.

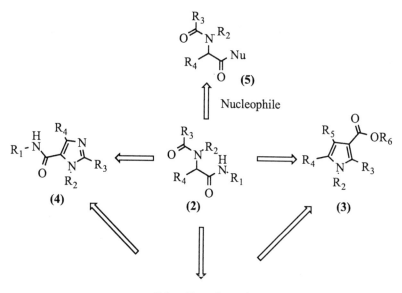

Figure 2. N-alkyl-N-acyl-α-amino Amide Transformations

Linkers on Solid Support. The synthesis of N-alkyl-N-acyl-α-amino amide based libraries requires the attachment of one of the reactants (isocyanide, aldehyde, amine or acid) onto solid support. Examples of aldehyde, amine or isocyanide components are attached to polymer via ester, amide or ether linkages are described below. For example, the reaction of Wang resin with $HOOC(CH_2)_nNHCHO$ (n= 2-10) in the presence of DIC, DMAP and triethylamine afforded the formamide **6** in excellent yield. The amide was converted to the corresponding isocyanide **7** using triphenylphosphine and carbon tetrachloride in the presence of triethylamine (99%) (Figure 3).

In an example, the Fmoc-protected amine linker $HOOC(CH_2)_nNHFmoc$ was attached to Rink resin using HBTU, HOBT in DMF to give the amide **8** which, upon deprotection with 20% piperdine in DMF, afforded the desired amine **9** was obtained in high yield (99%) (Figure 4) (*8*).

Mitsunobu reaction (*9*) between a series of phenols and an extended benzyl alcohol linker-bound to solid support has been reported (*10*). Mitsunobu reaction between phenols and polystyrene based polymers bearing benzyl alcohols as linkers has not been exploited on solid support. We have found that electron rich phenols such as p-aminophenol undergoes Mitsunobu reaction with Wang resin using azadicarbonyl dipiperidine (ADDP) and tri-n-butylphosphine in 95% yield (Figure 5).

However, electron poor phenols such as p-hydroxybenzaldehyde undergoes Mitsunobu reaction in poor yield (< 20%). The di-aza compound **12** was formed instead as the major product (> 67 %) (*11*). An alternative method was to convert the

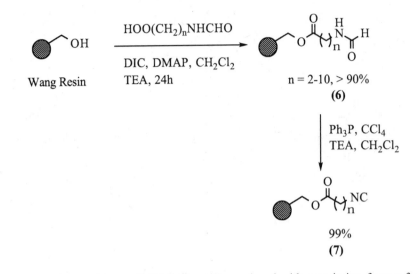

Figure 3. Isocyanides on Solid Support (Reproduced with permission from ref. 6. Copyright 1996 Elsevier Science Ltd.)

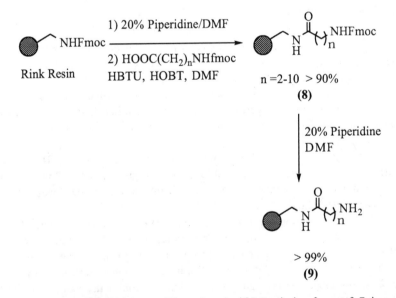

Figure 4. Amine on Solid Support (Reproduced with permission from ref. 7, in press. Copyright 1996 Elsevier Science Ltd.)

Wang resin-OH to the corresponding Wang resin-Br **13** using CBr₄ and triphenylphosphine in the presence of DMF. Reaction of p-hydroxybenzaldehyde with resin **13** in the presence of triethyl amine and DMF provided the desired product **14** in good yield (67%) (Figure 6).

Convertible Isocyanide in a Four Component Condensation Reaction. In general 2° amides are difficult to hydrolyze to their corresponding acids or esters (*12*). However, examples of 2° amides such as N-t-Boc-benzyl amide derivatives (*13*), 1-cyclohexenyl amides (*14*) and 2-azidophenyl amides (*15*) may be converted to their corresponding acids or esters. The need for a catalytic hydrogenation step during the latter method limits the scope of this reaction.

An ideal isocyanide would be easy to synthesize, readily undergo 4-component condensation reactions and offer facile conversion to other derivatives such as acids, esters, amides, ureas, etc. Initial experiments showed that BnNC reacts with a variety of aldehydes, acids and amines to provide the corresponding N-acyl-N-alkyl-α-aminobenzyl amides. These could be treated with t-Boc₂O-DMAP in THF followed by LiOH hydrolysis in a 1:1 THF-water mixture to give the corresponding N-acyl-N-alky-α-amino acids. Unfortunately both reactions were very slow (t₁/₂>7days) with low overall yield (10%). The rate and the yield of these reactions improved drastically when BnNC was replaced with PhNC.

Reaction of the amine-pound polymer **9** (Figure 7) with a series of aldehydes (R₁CHO), acids (R₂COOH) and Phenyl isocyanide in a 1:1:1 mixture of MeOH/CHCl₃/Pyridine at 65°C provided the corresponding Ugi product **15** attached to the polymer in good yield (50-70%). We observed no reaction at room temperature. We also observed that pyridine was essential to stabilize the isocyanide. Phenyl amides **15** react with t-Boc₂O and DMAP in the presence of TEA to provide amide **16** in excellent yield (> 95%). Compounds **16** were hydrolyzed to the corresponding carboxylic acids **17** using basic conditions. Cleavage from the polymer was carried out using 10% trifluoroacetic acid in dichloromethane providing the acids **18** in quantitative yield.

Post Four Component Condensation Reaction Transformations: 5-Membered Heterocycle Synthesis

Penta-substituted Pyrroles. Acids **18** (Figure 8) were subjected to neat acetic anhydride or isobutyl chloroformate and triethylamine in toluene followed by the addition of a series of acetylenic esters to provide polymer bound penta-substituted pyrroles **20**. The reaction proceeds via formation of the münchnone **19** and the *in situ* [3 + 2] cycloaddition with a variety of alkynes to yield the pyrroles **20**. Subsequent cleavage of pyrroles from solid support was accomplished using 20% TFA/CH₂Cl₂ to provide the pyrroles **21** and **22** as a mixture of isomers in approximately 4:1 ratio, respectively. Final products were obtained in good overall (yield is 35-72% over eight steps). This reflects a process in which each synthetic transformation occurred in an average of greater than 85% yield.

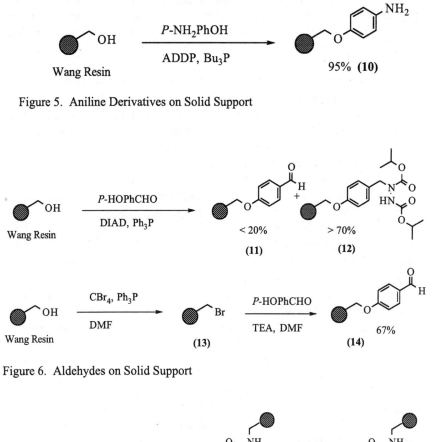

Figure 5. Aniline Derivatives on Solid Support

Figure 6. Aldehydes on Solid Support

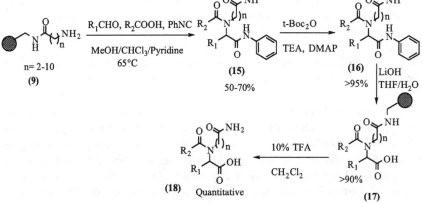

Figure 7   Phenyl Isocyanide in a Four Component Reaction on Solid Support
(Reproduced with permission from ref. 7, in press. Copyright 1996 Elsevier Science
Ltd.)

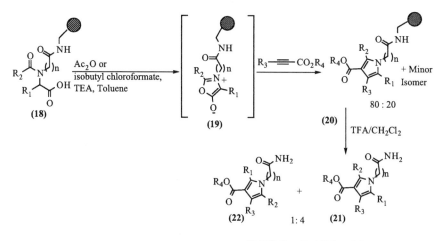

Figure 8. Pyrrole Formation on Solid Support (Reproduced with permission from ref. 7, in press. Copyright 1996 Elsevier Science Ltd.)

Some of the compounds prepared by this method are listed in Table I:

Table I.

n	R_1	R_2	R_3	R_4	%Yield
1	Et	4-Br-C_6H_4	CO_2Me	CO_2Me	46
2	Et	4-Br-C_6H_4	CO_2H	CO_2H	49
2	Et	4-Br-C_6H_4	Et	CO_2Et	35
2	i-Pr	$PhCH_2$	CO_2Me	CO_2Me	72
2	n-Pr	Ph	CO_2Me	CO_2Me	46
2	i-Pr	4-MeO-C_6H_4	CO_2Me	CO_2Me	45
2	n-Bu	4-CF_3-C_6H_4	CO_2Me	CO_2Me	40

The regiochemistry of these isomers was assigned using nOe experiments as shown in Figure 9. This ratio of cycloaddition is consistent with results obtained via solution chemistry (*16*).

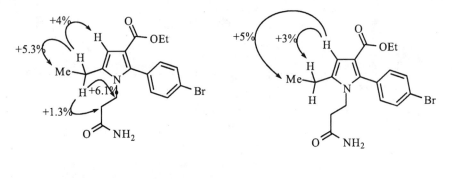

Major   Major

Figure 9. Regiochemistry Determination by nOe Experiments

Tetra-Substituted Imidazoles.   It has been reported that N-alkyl-N-acyl-α-aminoketones react with ammonium acetate in acetic acid at 100°C to provide the corresponding imidazoles (*17*).   Use of α-keto aldehydes in a 4-component condensation reaction (described above) would result in the synthesis of N-alkyl-N-acyl-α-aminoketone based libraries.   These ketone derivatives would then be converted to highly substituted imidazoles with similar diversity discussed earlier in this paper.   The isocyanide **7** was reacted with a series of aromatic ketoaldehydes, amines and carboxylic acids to provide the corresponding polymer-bound keto di-amides **23**.   These were then heated with ammonium acetate in acetic acid at 100°C to afford the corresponding tetra-substituted imidazoles **24** attached to the polymer. Compound **24** was treated with 10% trifluoroacetic acid in dichloromethane to provide the desired substituted imidazoles **25** in good overall yield (44-56%).   In addition, cleavage of **23** with 10% trifluoroacetic acid in dichloromethane gave the aryl ketones **26** in good overall yield (45-60%).

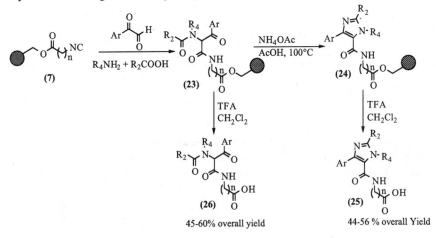

Figure 10. Imidazole Formation on Solid Support (Adapted from ref. 6.)

Examples of imidazoles (Ar = 4-X-C_6H_4) synthesized by this method are shown in Table II.

Table II.

n	X	R_2	R_4	%Yield
10	H	Ph	i-C_4H_9	45
10	H	4-F-Ph	i-C_4H_9	35
10	H	$PhCH_2$	i-C_4H_9	47
10	H	Ph	4-MeO-Ph	43
10	MeO	Ph	i-C_4H_9	51
2	H	Ph	i-C_4H_9	44
10	F	Ph	i-C_4H_9	49

Conclusion

Experimental library diversity (practical synthesis) relies entirely upon the availability of optimized synthetic methods which can then be carried out using automated processes. Such processes must encompass both software (library planning, compound registration, compound characterization, bioactivity analysis) and hardware which is capable of performing reactions over a wide range of experimental conditions to produce SDCLs.

The process described above demonstrates the general applicability of organic synthesis on solid support to drug discovery. One of the outstanding benefits is that several chemical transformations can be carried out without the need for chromatographic purification between each step. Furthermore, high speed synthesis of significant amounts of highly functionalized pharmacophores of high purity, coupled with high throughput screening, clearly accelerates the discovery and optimization of lead compounds.

Literature Cited

1. (a) Cao, X.; Siev, D.; Moran, E. J.; Lio, A.; Ohashi, C.; Mjalli, A. M. M. *Bioorganic & Med. Chem. Let.* **1995**, *5*, pp. 2953, (b) Baldwin, J. J.; Bourbon, J. J.; Henderson, I.; Ohlmeyer, M. H. J. *J. Am. Chem. Soc.* **1995**, *117*, pp. 5588-5589, (c) Zucherman, R. N.; Martin E. J.; Spellmeyer, D. C, et al., *J. Med. Chem.* **1994**, *37*, pp. 2678-85, (d) Moran, E. J.; Sarshar, S.; Cargill, J. F.; Shahbaz; M.; Lio, A.; Mjalli, A. M. M.; Armstrong, R. W. *J. Am. Chem. Soc.* **1995**, *117*, pp. 10787 and (e) Gallop, M. A.; Barrett, R. W.; Dower, W. J.; Fodor, S. P. A.; Gorden, E. M. *J. Med. Chem.* **1994**, *37*, pp. 1233-1251 and pp. 1385-1401.

2. (a) Mjalli, A. M. M.; Toyonaga, B. E. *Network   Science* **1995**, *1*, http://www.awod.com/netsci and (b) Nicolaou, K. C.; Xiao, Y.; Parandoosh, Z.; Senyei, A. and Nove, M. P. *Angew. Chem. Int. Ed. Eng.* **1995**, *107*, pp. 2476.

80   MOLECULAR DIVERSITY AND COMBINATORIAL CHEMISTRY

3. (a) Armstrong, R. W. International Patent WO 95192566, **1995** and (b) Bunin, B. A.; Plunkett, M. J.; Ellman, J. A. *Proc. Natl. Acad. Sci. USA* **1994**, *91*, pp. 4708 and references therein.

4. (a) Martin, E. J.; Blaney, J. M.; Siani, M. A.; Spellmeyer, D. C.; Wong, A. K.; Moos, W. H. *J. Med. Chem.* **1995**, *38*, pp. 1431-1436, (b) Cramer, R. D.; Patterson, D. E.; Bunce, J. D *J. Am. Chem. Soc.* **1988**, *110*, pp. 5959-596 and (c) Cramer, R. D.; Johnes, D. M; Patterson, D. E.; Simeroth, P. E. *Tetrahed. Comp. Meth.*, **1990**, *3*, pp. 47-59.

5. Ugi, I. *Isonitrile Chemistry*, Blomquist, A. T., Ed.; Academic Press: New York, **1971**, pp. 133.

6. Zhang, C. Z; Moran, E. J.; Woiwode, T. F.; Short, K. M.; Mjalli, A. M. M. *Tetrahedron Let.*, **1996**, *6*, 751.

7. Mjalli, A. M. M.; Sarshar, S.; Baiga, T. J. *Tetrahedron Let.*, **1996**, in press.

8. (a) Knorr, R.; Trzeciak, A.; Bannwarth, W.; Gillessen, D. *Tetrahedron Lett.* **1987**, *30*, pp. 1927 (b) Dourtoglow, V.; Gross, B. *Synthesis* **1984** pp. 572.

9. Hughes, D. L. *Org. Reactions* **1992**, *42*, pp. 335.

10. Rano, T.; Chapman, K. T. *Tetrahedron Lett.* **1995**, *36*, pp. 3789.

11. Tsunoda T., Yamamiya Y., Ito, S. *Tetrahedron Let.* **1993**, *34*, pp. 1639.

12. Corey, E. J.; Letavic, M. A. *J. Am. Chem. Soc.* **1995**, *117*, pp. 9616.

13. Griego, P. A. et al. *J. Org. Chem.* **1983**, *48*, pp. 2424.

14. Keating, T. A. et al. *J. Am. Chem. Soc.*, **1995**, *117*, pp. 7842.

15. Ugi, I. *Angew. Chem. Int. Ed. Engl.* **1982**, *21*, pp. 810.

16. Coppola, B. P.; Noe, M. C.; Schwartz, D. J.; Abdon, R. L. II; Trost, B. M. *Tetrahedron Lett.* **1994**, *50*, pp. 93.

17. Baumgartel, H., et al. *Chem Ber.* **1973**, *106*, pp. 2415.

Chapter 9

A Solution-Phase Strategy for the Parallel Synthesis of Chemical Libraries Containing Small Organic Molecules

A General Dipeptide Mimetic and a Flexible General Template

Christine M. Tarby[1], Soan Cheng[1], and Dale L. Boger[2]

[1]CombiChem, Inc., 9050 Camino Sante Fe, San Diego, CA 92121
[2]Department of Chemistry, Scripps Research Institute,
10666 North Torrey Pines Road, La Jolla, CA 92037

Abstract. A general approach to the solution phase, parallel synthesis of chemical libraries, which allows the preparation of multi-milligram quantities of each individual member, is exemplified with both a dipeptide mimetic and flexible general template. In each step of the sequence, the reactants, unreacted starting material, reagents and their byproducts are removed by simple liquid/liquid or liquid/solid extractions providing the desired intermediates and final compounds in high purities (≥90–100%) independent of the reaction yields and without deliberate reaction optimization.

Given its ability to rapidly and efficiently produce large numbers of diverse compounds in a cost-effective manner in conjunction with high-throughput screening of the ever increasing number of molecular targets, combinatorial synthesis has emerged as a powerful tool for the acceleration of the drug discovery process (1-5). The implications of the technology are apparant arising from both the production of lead-generation libraries comprised of large numbers of diverse compounds or from the production of smaller targeted libraries for optimization around a promising lead candidate. Initially explored with peptide or oligonucleotide libraries and related structures (2-3,6-21), more recent efforts have been directed at exploiting the diversity and range of useful properties embodied in conventional small molecule synthesis of predominantly "drug-like" molecules (16-44). A variety of solid phase methods have been utilized for the generation of chemical libraries including split and mixed (45-48), encoded (49-57), indexed (58-60), or parallel and spatially addressed synthesis on pins (6, 27), beads (61), chips (62) and other solid supports (63-70) while solution phase combinatorial synthesis has not been embraced as a viable alternative (58-60, 71). This may be attributed to the natural evolution of

1054–7487/96/0081$15.00/0
© 1996 American Chemical Society

combinatorial chemistry from solid phase peptide and oligonucleotide synthesis where supported phase synthesis has emerged as the method of choice for automated repetitive coupling reactions. Solid phase synthesis offers the two advantages of product isolation and sample manipulation which are important issues for library synthesis. Attaching a substrate to a resin allows for product isolation by simple filtration and permits the use of large reagent excesses to effect high yield conversions required for each of the steps (*72*). While the limitations of solid phase synthesis have been addressed for the repetitive reactions used in the production of peptides and oligonucleotides, the disadvantages for small molecule synthesis where each sequential reaction is potentially different are well recognized (*72*). The scale is restricted by the required amount of solid support and its loading capacity. The production of multi-milligram quantities of each member can be cumbersome and potentially prohibitively expensive for large or even medium sized libraries (1000–5000 members) (*73*). Supported phase synthesis also requires functionalized substrates and solid supports (*63-70*), compatible spacer linkers, and orthogonal chemistries for attachment and detachment which can often yield undesired functional groups at the detachment point. In many instances, due to the limited repertoire of chemical reactions available on solid phase, the chemistry must be optimized not only for the substrate but for the resin as well. This is made more challenging by the fact that many of these resins have been developed and optimized for solid phase peptide and oligonucleotide synthesis and may not be ideal for a diverse range of chemistries. It also requires the use of specialized analytical techniques for monitoring the individual steps of a multistep synthesis (*74-82*) as well as orthogonal capping strategies for blocking unreacted substrate and does not permit for the purification of resin bound intermediates. This latter feature becomes most significant when dealing with non-peptides/peptoids and necessarily produces the released product of a multistep sequence in an impure state unless each reaction on the substrate proceeds with unusually high efficiency. Like the efforts on even the repetitive steps of the solid phase synthesis of peptides or oligonucleotides, the optimization of the reactions for assuring the required reaction efficiencies is time consuming and challenging. For even a modest criterion of final product purity (at least 85% pure), this requires that each step of even a three-step reaction sequence proceed in 95% yield on each substrate. Our experience has been that such generalized reaction efficiencies with a wider range of chemistries and substrates are not routinely obtainable and require such an extensive investment in reaction optimization and purification that the overall efficiency of the combinatorial technique is compromised. [Representative examples include a two-step solid phase synthesis of HIV protease inhibitors utilizing two sequential amide couplings provided the final released agents in purities ranging from 30–70% or 20–50% (*83*).]

Consequently, an important complement to adapting solution phase chemistry to solid phase combinatorial synthesis is the development of protocols for solution phase combinatorial synthesis. Given that solution and solid-phase sample manipulation are both convenient and easily automated, the only limitation to the solution phase parallel synthesis of chemical libraries is the *isolation* of reaction products away from the reagents and by products. If the advantages of sample isolation attributed to a solid phase synthesis may be embodied in a solution phase synthesis, its non-limiting scale, expanded and non-limiting repertoire of chemical reactions, direct production of soluble intermediates and final products for assay or for purification, and the avoidance of linking and capping strategies make solution phase combinatorial synthesis an attractive alternative. A number of techniques are available for product isolation and one of the most simple and attractive is liquid/liquid or solid/liquid extraction. Herein, we provide details of high purity solution phase parallel syntheses of chemical libraries around two templates which implement simple isolation and purification protocols at each step (*84-86*).

A Generalized Dipeptidomimetic Template

Template **1** is a designed rigid core structure which contains a number of important features. When fully extended, the template contains a rigid bicyclic core with a plane of symmetry which enables it to function as a Gly-X mimic (Figure 1). When positions 1 and 3 are extended, the conformation mirrors that of an extended sheet as shown by the superimposition of extended **1** with an simple peptide (H$_2$N-Ala-Gly-Ala-CONHCH$_3$) in Figure 2. Extension of positions 1 and 2 introduce a turn motif. When all three positions are utilized, an interesting core peptidomimetic which explores three-dimensional space is produced. Its symmetrical structure contains three positions which can be controllably and sequentially functionalized with a variety of nucleophiles and acylating agents enabling the synthesis of libraries with three variable units. Importantly, at each step of the synthesis, a handle is released which allows not only for the isolation of the desired product but its purification from reaction byproducts and reagents as well.

As a five-membered cyclic anhydride, the starting template is activated for the first functionalization which upon reaction liberates a carboxylic acid, its second functionalization site. As such, no orthogonal protecting groups are required for the template functionalization and four chemical steps are required for N^3 diversification. The same released functionality (CO$_2$H, NH) may be used as a handle for isolation and purification of the expected products from starting materials, reagents and reaction byproducts by simple liquid/liquid or solid/liquid extraction. Any nucleophile can be added to open the starting template anhydride. Following functionalization of the released acid, removal of an orthogonal protecting group on nitrogen allows an additional stage for purification and a subsequent acylating agent to be added to complete the diversification. In addition to the use of an amine in the second diversification step detailed herein, this second step has also been modified to accommodate the use of any neutral nucleophile (R^{2}OH, R^{2}SH or nucleophile) by conducting an additional extraction purification on the amine liberated in the subsequent *N*-BOC deprotection. In each step of the sequence, the reactants, unreacted starting materials, reagents and their byproducts can be removed by simple extractions, providing the intermediates and final compounds in high purities.

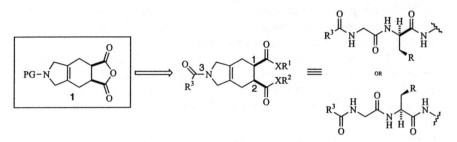

Figure 1. A General Dipeptide Mimetic.

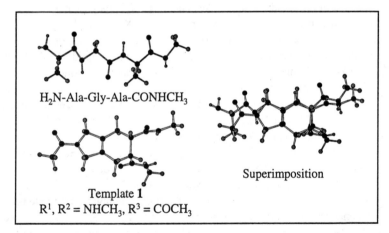

Figure 2. The superimposition of extended **1** with a simple peptide (H$_2$N-Ala-Gly-Ala-CONHCH$_3$).

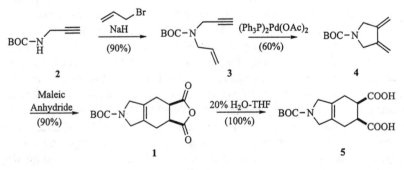

Scheme 1

The template synthesis (Scheme 1) requires N-BOC protection of propargyl amine and subsequent alkylation effected by treatment with NaH (1.1 equiv, DMF, 25 °C, 30 min) followed by allyl bromide (1.2 equiv, 0 °C, 5 h) to generate **3** (>90% yield, two steps). Treatment of **3** with catalytic (Ph$_3$P)$_2$Pd(OAc)$_2$ (0.5 equiv, 80 °C, C$_6$H$_6$, 1 h) affords diene **3** (60%) (*87*). The reactive diene is immediately subjected to a Diels-Alder reaction with maleic anhydride (1 equiv, C$_6$H$_6$, 40 °C, 1 h) to yield **1** which upon deliberate hydrolysis (20% H$_2$O-THF, 5 h) provides the easily purified and handled diacid **5**. The anhydride **1** is regenerated *in situ* upon treatment with EDCI (1 equiv) immediately prior to the addition of the first nucleophile.

To illustrate the library construction with **1** which is representative of efforts with related five-membered cyclic anhydrides, we have recently disclosed details of our initial efforts which were conducted without optimization and that provided a fully characterized 27 member library constructed as a 3 x 3 x 3 matrix yielding 39 unique components in individual vessels constituting a library of 78 compounds including enantiomers (Figures 3 and 4) (*84,86*). Treatment of **5** with EDCI (1.1 equiv, DMF, 25 °C 20 min) followed by addition of R^{1}NH$_2$ (1 equiv, 25 °C, 16 h) afforded the monoamides which were purified by simple acid/base dissolution (80–99%). Importantly, only the monoamide product was generated indicating *in situ* closure of the initially generated activated carboxylate to the anhydride **1** and its subsequent reaction with the added amine. The monoamides were split into four equal components with one being retained for archival purposes. Each of the three remaining aliquots were treated with excess reagent EDCI (3 equiv) and R^{2}NH$_2$ (3 equiv, DMF, 25 °C, 16 h) to yield 9 diamides (65–91%) which were purified by an acid and base washing removing the excess unreacted reactants, reagents and reagent byproducts. Although this has been represented in Figure 3 with aliphatic amines, primary aryl amines also react well. One-quarter of the diamide was retained and the remaining quantity was subjected to N-BOC deprotection (4 N HCl-EtOAc, 25 °C, 30 min). One-third of each was treated with EDCI (2 equiv) and R^{3}CO$_2$H (2 equiv, DMF, 25 °C) such that 27 unique products were obtained. The resulting functionalized peptidomimetics were purified by washing with aqueous acid and base to yield the purified final compounds (3–89%). Irrespective of individual yields, the intermediates were 95-100% pure and final compounds were ≥90–95% pure. The only contaminant occasionally observed was a small quantity of the oxidized pyrrole which was minimized by the careful exclusion of oxygen during the N-BOC deprotection and subsequent acylation. During the course of the reaction sequence, no specialized monitoring protocols were required and reactions were observed by TLC. Additionally, each individual compound synthesized (intermediate and final product) was traditionally characterized by ^{1}H NMR and high resolution FABMS.

The Flexible Iminodiacetic Acid Template.

The template **6**, which is representative of a set of six-membered cyclic anhydride-based templates that have been examined, consists of a densely functionalized core

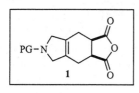

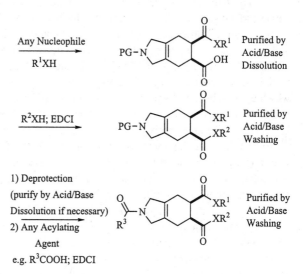

Figure 3. The extension, isolation and purification of template **1**.

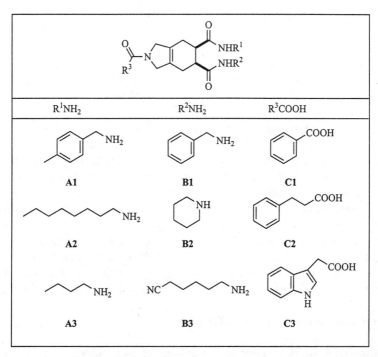

Figure 4. 3 x 3 x 3 Library based on template **1**.

which imposes little structural or conformational bias which might limit its use. The added pendant groups provide the molecular diversity and libraries built upon **6** may prove applicable to many biological targets. Its symmetrical structure contains three positions which can be sequentially functionalized enabling the synthesis of libraries with up to three variable units (Figure 5). Like **1,** the template is activated for the first functionalization with *in situ* generation of the anhydride which upon reaction and opening liberates a free carboxylic acid as its second functionalization site. Thus, no orthogonal protecting groups are required for the template functionalization and only four chemical steps are required for the N^3 diversification (Scheme 2). Also like **1,** the same released functionality at each step may be used for both the *isolation and purification* of each intermediate and final product from the starting material, reactants, reagents and their reaction byproducts by simple liquid/liquid or solid/liquid extraction providing highly pure materials (≥90–100%) independent of the reaction efficiencies.

In our recent disclosure (85, *86*), we described full details of our initial efforts with **6** which were conducted without prior optimization and that provided a fully characterized 27 member library constructed as a 3 x 3 x 3 matrix affording 39 unique components in individual vessels (Scheme 2, Figure 6). Although this has been subsequently expanded to much larger libraries, the disclosure of its initial implementation served to highlight a number of the advantages of the approach. Each of the expected library members was obtained in a purified form (≥90–100% pure) independent of the reaction efficiencies in amounts ranging from 5–60 mg. *In situ* closure of *N*-BOC-iminodiacetic acid to the anhydride **6** (1 equiv EDCI, DMF, 25 °C, 1 h) followed by treatment with one of three R^1NH_2 (1 equiv, DMF, 25 °C, 20 h, 84–86%) cleanly afforded the monoamides which were purified by simple acid extraction to remove unreacted R^1NH_2, EDCI, and its reaction byproducts. The three monoamides were each partitioned into three portions with one smaller portion being retained for archival purposes. Each of the equal three portions were treated with three R^2NH_2 (1 equiv) and PyBOP (1 equiv, 2 equiv *i*-Pr$_2$NEt, DMF, 20 °C, 25 h, 65–99%) to afford nine diamides which were effectively purified by acid and base extractions to removed reaction byproducts, the unreacted starting material and R^2NH_2, PyBOP, and its reaction byproducts. Although this is illustrated in Scheme 2 with primary amines, secondary amines as well as aryl amines have been found to work as well. Following the second functionalization and *N*-BOC deprotection (4 N HCl, dioxane, 25 °C, 45 min), reaction of three equal portions of each amine with three R^3CO_2H (1 equiv) in the presence of PyBOP (1 equiv, 3 equiv *i*-Pr$_2$NEt, DMF, 25 °C, 20 h, 16–100%) provided 27 agents which were purified by aqueous acid and base extractions to remove unreacted starting materials, reagents, and their reaction byproducts.

Overall yields for the 27 agents ranged from 9–84% with an average overall yield of 61% for the three derivatizations. During the course of the reaction sequence, no specialized monitoring protocols were required and reactions were observed by TLC. Additionally, each individual compound synthesized (intermediate and final

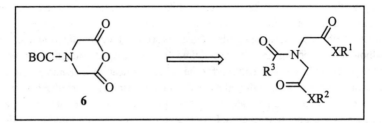

Figure 5. The Iminodiacetic Anhydride Template.

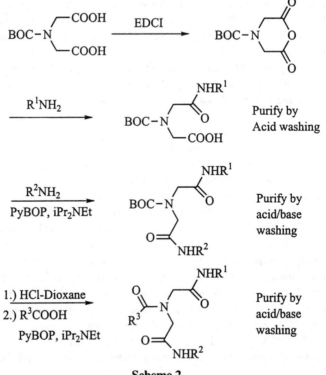

Scheme 2

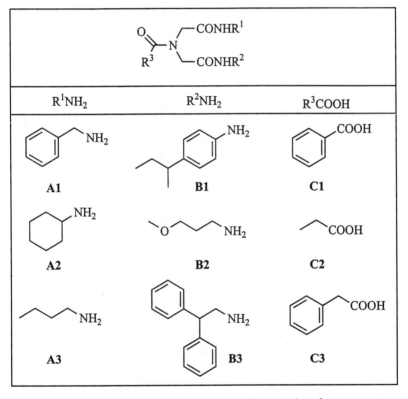

Figure 6. 3 x 3 x 3 Library based on template **6**.

product) was traditionally characterized by ^{1}H NMR and high resolution FABMS. Importantly, and independent of individual yields, all intermediates and final products were ≥90% pure with an average 95% purity as determined by HPLC analysis (Table I). Without optimization in these first efforts, most of the final library products were obtained in 20–60 mg quantities as individual samples at this exceptional level of purity suitable for direct use in screening efforts without further purification.

Table I. HPLC Purity of the 3 x 3 x 3 Library Intermediates and Final Products

Agent	Purity (%)	Agent	Purity
A1	100	A1B3C3	95.6
A2	95.6	A2B1C1	100
A3	100	A2B1C2	90.1
A1B1	100	A2B1C3	95.2
A1B2	95.6	A2B2C1	93.6
A1B3	94.8	A2B2C2	nd*
A2B1	96.5	A2B2C3	95.6
A2B2	97.5	A2B3C1	92.6
A2B3	100	A2B3C2	91.3
A3B1	100	A2B3C3	93.7
A3B2	nd*	A3B1C1	97.5
A3B3	97.5	A3B1C2	96.1
A1B1C1	94.0	A3B1C3	96.8
A1B1C2	91.3	A3B2C1	94.3
A1B1C3	96.2	A3B2C2	nd*
A1B2C1	93.5	A3B2C3	96.7
A1B2C2	95.3	A3B3C1	90.7
A1B2C3	90.6	A3B3C2	96.1
A1B3C1	91.2	A3B3C3	91.4
A1B3C2	92.9		

*Not determined, no UV active chromophore

Subsequent extensions of these efforts to the preparation of a 125 member library constructed as a 5 x 5 x 5 matrix afforded 155 unique compounds (Figure 7) and provided comparable observations. In this example, the amines and carboxylic acids used were selected from our medicinal chemistry monomer sets. Each library member was obtained as an individual entity in 30–100 mg quantities in pure form (>90%, generally ≥95% pure) in overall yields ranging from 32–85% (64% average). The library was traditionally characterized along the matrix by ^{1}H NMR and high resolution FABMS to insure each reaction type was successful and every

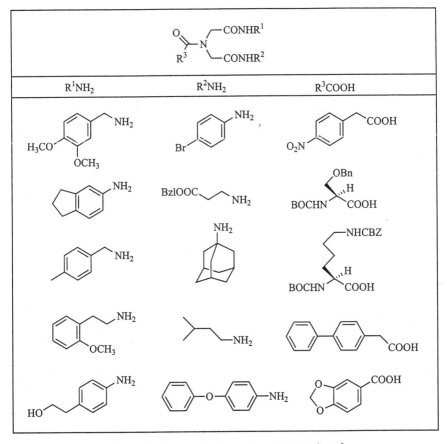

Figure 7. 5 x 5 x 5 Library based on template **6**.

intermediate and final product in the library was confirmed by Electrospray
Ionization MS.

One of the largest libraries based on **6** addressed to date by manual
manipulation of the reactions was a 960 member library constructed in a 6 x 8 x 20
matrix affording 1014 final components in individual vessels including intermediates
constituting a library of 1158 compounds including diastereomers and enantiomers
(Figure 8). Each library member was obtained in amounts ranging from 30–150 mg in
yields ranging from 10–71% with an average overall yield of 52%.

Discussion and Conclusions

Complementary solid phase combinatorial library synthesis, a solution phase method
has been developed for the rapid and simple multistep, parallel synthesis of chemical
libraries in which each component is produced as an individual compound in
potentially unlimited quantities (typically 30–150 mg) in a format directly compatible
with most screening assays. In each step of the sequence, the intermediates and all
final products were subjected to a simple isolation and purification protocol in which
liquid/liquid or liquid/solid extraction is used to remove reactants, unreacted starting
material, reagents, and their byproducts providing the library members in high purities
(≥90–100%) irrespective of the individual reaction yields and without deliberate
reaction optimization. The template **6** is unusually flexible possessing 1–3
functionalization sites for diversification and little inherent structural or
conformational bias which might limit its use as a general template. A global diversity
measure of the libraries based on **6** are currently under evaluation by our informatics
team such that upon addition to and redesign of these libraries, the maximum diversity
level obtainable within the scaffold constraints is obtained. The five-membered cyclic
anhydride **1** constitutes a rigid dipeptidomimetic template capable of serving as a
nonpeptide scaffold for either extended or turn peptidomimetics and possesses 1–3
functionalization sites. It, along with other related rigid templates, should be useful
not only as lead generation libraries in their own right but as secondary libraries for
further optimization and exploration of leads discovered with the more flexible
template **6**. Although the solution phase technology detailed herein enlists
conventional liquid/liquid extractions, similar results employing solid-supported
resins, columns, or pads have been used to effect solid/liquid extractions by simple
batch, column, or filtration protocols. The one secondary amine protecting group
may be easily altered to accommodate its sensitivity to selected liquid/liquid or
liquid/solid extraction protocols used to remove starting materials and reaction
byproducts. In addition, the approach is not limited to amide bond forming reactions.
Other nucleophiles may be utilized in the first functionalization of the anhydride
templates with purification of the desired product by dissolution in base. Similarly,
the second functionalization may be accomplished by reaction of the activated
carboxylate with other nucleophiles followed by purification of the desired product

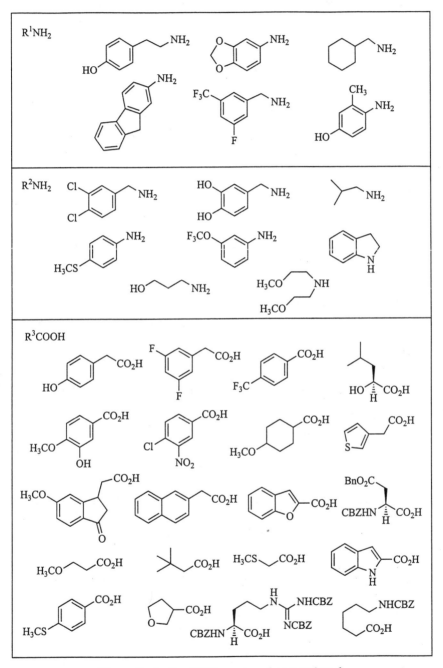

Figure 8. 6 x 8 x 20 Library based on template **6**.

by dissolution in aqueous acid following *N*-BOC deprotection. Although not illustrated herein, the strategy is also not limited to the parallel synthesis of individual compounds but is also applicable to split and mixed synthesis employing stoichiometric limiting variable units and excess template to construct combinatorial libraries of compound mixtures subject to subsequent compound identification by repeat synthesis or recursive deconvolution. Studies employing larger targeted libraries with matrix characterization of each reaction type, their adaptation to automation, the development of related library templates, as well as additional approaches to the solution phase synthesis of chemical libraries will be disclosed in due time.

Literature Cited

1.   Terrett, N. K.; Gardner, M.; Gordon, D. W.; Kobylecki, R. J.; Steele, J. *Tetrahedron* **1995**, *51*, 8135.
2.   Gallop, M. A.; Barrett, R. W.; Dower, W. J.; Fodor, S. P. A.; Gordon, E. M. *J. Med. Chem.* **1994**, *37*, 1233.
3.   Gordon, E. M.; Barrett, R. W.; Dower, W. J.; Fodor, S. P. A.; Gallop, M. A. *J. Med. Chem.* **1994**, *37*, 1385.
4.   Janda, K. D. *Proc. Natl. Acad. Sci. U.S.A.* **1994**, *91*, 10779.
5.   Pavia, M. R.; Sawyer, T. K.; Moos, W. H. *Bioorg. Med. Chem. Lett.* **1993**, *3*, 387.
6.   Geysen, H. M.; Meloen, R. H.; Barteling, S. J. *Proc. Natl. Acad. Sci. U.S.A.* **1984**, *81*, 3998.
7.   Lam, K. S.; Salmon, S. E.; Hersh, E. M.; Hruby, V. J.; Kazmierski, W. M.; Knapp, R. J. *Nature* **1991**, *354*, 82.
8.   Houghten, R. A.; Pinilla, C.; Blondelle, S. E.; Appel, J. R.; Dooley, C. T.; Cuervo, J. H. *Nature* **1991**, *354*, 84.
9.   Salmon, S. E.; Lam, K. S.; Lebl, M.; Kandola, A.; Khattri, P. S.; Wade, S.; Patek, M.; Kocis, P.; Krchnak, V.; Thorpe, D.; Felder, S. *Proc. Natl. Acad. Sci. U.S.A.* **1993**, *90*, 11708.
10.   Owens, R. A.; Gesellchen, P. D. Houchins, B. J.; DiMarchi, R. D. *Biochem. Biophys. Res. Commun.* **1991**, *181*, 402.
11.   Bock, L. C.; Griffin, L. C.; Latham, J. A.; Vermaas, E. H.; Toole, J. J. *Nature* **1992**, *355*, 564.
12.   Scott, J. K.; Smith, G. P. *Science* **1990**, *249*, 386.
13.   Cwirla, S. E.; Peters, E. A.; Barrett, R. W.; Dower, W. J. *Proc. Natl. Acad. Sci. U.S.A.* **1990**, *87*, 6378.
14.   Devlin, J. J.; Panganiban, L. C.; Devlin, P. E. *Science* **1990**, *249*, 404.
15.   Freier, S. M.; Konings, D. A. M.; Wyatt, J. R.; Ecker, D. J. *J. Med. Chem.* **1995**, *38*, 344.
16.   Simon, R. J.; Kania, R. S.; Zuckermann, R. N.; Huebner, V. D.; Jewell, D. A.; Banville, S.; Ng, S.; Wang, L.; Rosenberg, S.; Marlowe, C. K.; Spellmeyer, D.

C.; Tam, R.; Frankel, A. D.; Santi, D. V.; Cohen, F. E.; Bartlett, P. A. *Proc. Natl. Acad. Sci. U.S.A.* **1992**, *89*, 9367.

17. Zuckermann, R. N.; Kerr, J. M.; Kent, S. B. H.; Moos, W. H. *J. Am. Chem. Soc.* **1992**, *114*, 10646.

18. Miller, S. M.; Simon, R. J.; Ng, S.; Zuckermann, R. N.; Kerr, J. M.; Moos, W. H. *Bioorg. Med. Chem. Lett.* **1994**, *4*, 2657.

19. Zuckermann, R. N.; Martin, E. J.; Spellmeyer, D. C.; Stauber, G. B.; Shoemaker, K. R.; Kerr, J. M.; Figliozzi, G. M.; Goff, D. A.; Siani, M. A.; Simon, R. J.; Banville, S. C.; Brown, E. G.; Wang, L.; Richter, L. S.; Moos, W. H. *J. Med. Chem.* **1994**, *37*, 2678.

20. Cho, C. Y.; Moran, E. J.; Cherry, S. R.; Stephans, J. C.; Fodor, S. P. A.; Adams, C. L.; Sundaram, A.; Jacobs, J. W.; Schultz, P. G. *Science* **1993**, *261*, 1303.

21. Ostresh, J. M.; Husar, G. M.; Blondelle, S. E.; Dorner, B.; Weber, P. A.; Houghten, R. A. *Proc. Natl. Acad. Sci. U.S.A.* **1994**, *91*, 11138.

22. Bunin, B. A.; Ellman, J. A. *J. Am. Chem. Soc.* **1992**, *114*, 10997.

23. Bunin, B. A.; Plunkett, M. J.; Ellman, J. A. *Proc. Natl. Acad. Sci. U.S.A.* **1994**, *91*, 4708.

24. Virgilio, A. A.; Ellman, J. A. *J. Am. Chem. Soc.* **1994**, *116*, 11580.

25. Kick, E. K.; Ellman, J. A. *J. Med. Chem.* **1995**, *38*, 1427.

26. Boojamra, C. G.; Burow, K. M.; Ellman, J. A. *J. Org. Chem.* **1995**, *60*, 5742.

27. DeWitt, S. H.; Kiely, J. S.; Stankovic, C. J.; Schroeder, M. C.; Cody, D. M. R.; Pavia, M. R. *Proc. Natl. Acad. Sci. U.S.A.* **1993**, *90*, 6909.

28. Chen, C.; Randall, L. A. A.; Miller, R. B.; Jones, A. D.; Kurth, M. J. *J. Am. Chem. Soc.* **1994**, *116*, 2661.

29. Beebe, X.; Schore, N. E.; Kurth, M. J. *J. Am. Chem. Soc.* **1992**, *114*, 10061.

30. Moon, H.-S.; Schore, N. E.; Kurth, M. J. *Tetrahedron Lett.* **1994**, *35*, 8915.

31. Kurth, M. J.; Randall, L. A. A.; Chen, C.; Melander, C.; Miller, R. B.; McAlister, K.; Reitz, G.; Kang, R.; Nakatssu, T.; Green, C. *J. Org. Chem.* **1994**, *59*, 5862.

32. Gordon, D. W.; Steele, J. *Bioorg. Med. Chem. Lett.* **1995**, *5*, 47.

33. Patek, M.; Drake, B.; Lebl, M. *Tetrahedron Lett.* **1994**, *35*, 9169.

34. Patek, M.; Drake, B.; Lebl, M. *Tetrahedron Lett.* **1995**, *36*, 2227.

35. Campbell, D. A.; Bermak, J. C.; Burkoth, T. S.; Patel, D. V. *J. Am. Chem. Soc.* **1995**, *117*, 5381.

36. Forman, F. W.; Sucholeiki, I. *J. Org. Chem.* **1995**, *60*, 523.

37. Rano, T. A.; Chapman, K. T. *Tetrahedron Lett.* **1995**, *36*, 3789.

38. Dankwardt, S. M.; Newman, S. R.; Krstenansky, J. L. *Tetrahedron Lett.* **1995**, *36*, 4923.

39. Deprez, B.; Williard, X.; Bourel, L.; Coste, H.; Hyafil, F.; Tartar, A. *J. Am. Chem. Soc.* **1995**, *117*, 5405.

40. Murphy, M. M.; Schullek, J. R.; Gordon, E. M.; Gallop, M. A. *J. Am. Chem. Soc.* **1995**, *117*, 7029.

41. Kocis, P.; Issakova, O.; Sepetov, N. F.; Lebl, M. *Tetrahedron Lett.* **1995**, *37*, 6623.

42. Krehnak, V.; Flegelova, Z.; Weichsel, A. S.; Lebl, M. *Tetrahedron Lett.* **1995**, *37*, 6193.

43. Goff, D. A.; Zuckerman, R. N. *J. Org. Chem.* **1995**, *60*, 5748.

44. Terrett, N. K.; Bojanic, D.; Brown, D.; Bungay, P. J.; Gardner, M.; Gordon, D. W.; Mayers, C. J.; Steele, J. *Bioorg. Med. Chem. Lett.* **1995**, *5*, 917.

45. Furka, A.; Sebestyen, F.; Asgedom, M.; Dibo, G. *Abstr. 14th Intl. Congress Biochem., Prague* **1988**, *5*, 47.

46. Furka, A.; Sebestyren, F.; Asgedom, M.; Dibo, G. *Int. J. Peptide Protein Res.* **1991**, *37*, 487.

47. Houghten, R. A. *Proc. Natl. Acad. Sci. U.S.A.* **1985**, *82*, 5131.

48. Erb, E.; Janda, K. D.; Brenner, S. *Proc. Natl. Acad. Sci. U.S.A.* **1994**, *91*, 11422.

49. Brenner, S.; Lerner, R. A. *Proc. Natl. Acad. Sci. U.S.A.* **1992**, *89*, 5381.

50. Nielsen, J.; Brenner, S.; Janda, K. D. *J. Am. Chem. Soc.* **1993**, *115*, 9812.

51. Needels, M. C.; Jones, D. G.; Tate, E. H.; Heinkel, G. L.; Kochersperger, L. M.; Dower, W. J.; Barrett, R. W.; Gallop, M. A. *Proc. Natl. Acad. Sci. U.S.A.* **1993**, *90*, 10700.

52. Nikolaiev, V.; Stierandova, A.; Krchnak, V.; Seligmann, B.; Lam, K. S.; Salmon, S. E.; Lebl, M. *Peptide Res.* **1993**, *6*, 161.

53. Kerr, J. M.; Banville, S. C.; Zuckermann, R. N. *J. Am. Chem. Soc.* **1993**, *115*, 2529.

54. Ohlmeyer, M. H. J.; Swanson, R. N.; Dillard, L. W.; Reader, J. C.; Asouline, G.; Kobayashi, R.; Wigler, M.; Still, W. C. *Proc. Natl. Acad. Sci. U.S.A.* **1993**, *90*, 10922.

55. Nestler, H. P.; Bartlett, P. A.; Still, W. C. *J. Org. Chem.* **1994**, *59*, 4723. Baldwin, J. J.; Burbaum, J. J.; Henderson, I.; Ohlmeyer, M. H. J. *J. Am. Chem. Soc.* **1995**, *117*, 5588.

56. Moran, E. J.; Sarshar, S.; Cargill, J. F.; Shahbaz, M. M.; Lio, A.; Mjalli, A. M. M.; Armstrong, R. W. *J. Am. Chem. Soc.* **1995**, *117*, 10787.

57. Nicolaou, K. C.; Xiao, X.-Y.; Parandoosh, Z.; Senyei, A.; Nova, M. P. *Angew. Chem., Int. Ed. Eng.* **1995**, *34*, 2289.

58. Pirrung, M. C.; Chen, J. *J. Am. Chem. Soc.* **1995**, *117*, 1240.

59. Smith, P. W.; Lai, J. Y. Q.; Whittington, A. R.; Cox, B.; Houston, J. G.; Stylli, C. H.; Banks, M. N.; Tiller, P. R. *Bioorg. Med. Chem. Lett.* **1994**, *4*, 2821.

60. Carell, T.; Wintner, E. A.; Bashir-Hashemi, A.; Rebek, J., Jr. *Angew Chem., Int. Ed. Engl.* **1994**, *33*, 2059.

61. Merrifield, R. B. *J. Am. Chem. Soc.* **1963**, *85*, 2149.

62. Fodor, S. P. A.; Read, J. L.; Pirrung, M. C.; Stryer, L.; Lu, A. T.; Solas, D. *Science* **1991**, *251*, 767.

63. Atherton, E.; Sheppard, R. C. *Solid Phase Peptide Synthesis: A Practical Approach*; IRL Press: Oxford, 1989.

64. Grubler, G.; Stoeva, S.; Echner, H.; Voelter, W. in *Peptides: Chemistry, Structure, and Biology* (Proceedings of the Thirteenth American Peptide Symposium); Hodges, R. A.; Smith, J. A., Eds.; ESCOM-Leiden, The Netherlands, 1994, 51.
65. Englebretsen, D. R.; Harding, D. R. K. *Int. J. Peptide Protein Res.* **1992**, *40*, 487.
66. Frank, R. *Bioorg. Med. Chem. Lett.* **1993**, *3*, 425.
67 Frank, R.; Doring, R. *Tetrahedron* **1988**, *44*, 6031.
68. Schmidt. M.; Eichler, J.; Odarjuk, J.; Krause, E.; Beyermann, M.; Bienert, M. *Bioorg. Med. Chem. Lett.* **1993**, *3*, 441.
69. Eichler, J.; Bienert, M.; Stierandova, A.; Lebl, M. *Peptide Res.* **1991**, *4*, 296.
70. For traceless linkers, see: Plunkett, M. J.; Ellman, J. A. *J. Org. Chem.* **1995**, *60*, 6006.
71. For soluble polymer supports, see: Han, H.; Wolfe, M. M.; Brenner, S.; Janda, K. D. *Proc. Natl. Acad. Sci. U.S.A.* **1995**, *92*, 6419.
72. Crowley, J. I.; Rapoport, H. *Acc. Chem. Res.* **1976**, *9*, 135.
73. For a 10,000 member library (three-step synthesis at 95% yield/step) to obtain 50 mg of each component (MW = 500 g/mol) on Merrifield resin with a typical loading of 1 mmol/g requires 1.166 kg of solid support = $2499/library; on Wang resin with loading of 0.7 mmol/g requires 1.666 kg of solid support = $8331/library.
74. Egner, B. J.; Langley, G. J.; Bradley, M. *J. Org. Chem.* **1995**, *60*, 2652.
75. Anderson, R. C.; Jarema, M. A.; Shapiro, M. J.; Stokes, J. P.; Ziliox, M. *J. Org. Chem.* **1995**, *60*, 2650.
76. Fitch, W. L.; Detre, G.; Holmes, C. P.; Shoolery, J. N.; Kiefer, P. A. *J. Org. Chem.* **1994**, *59*, 7955.
77. Look, G. C.; Holmes, C. P.; Chinn, J. P.; Gallop, M. A. *J. Org. Chem.* **1994**, *59*, 7588.
78. Metzger, J. W.; Wiesmuller, K.-H.; Gnau, V.; Brunjes, J.; Jung, G. *Angew. Chem., Int. Ed. Engl.* **1993**, *32*, 894.
79. Youngquist, R. S.; Fuentes, G. R.; Lacey, M. P.; Keough, T. *Rapid Commun. Mass Spect.* **1994**, *8*, 77.
80. Chu, Y.-H.; Kirby, D. P.; Karger, B. L. *J. Am. Chem. Soc.* **1995**, *117*, 5419.
81. Brummel, C. L.; Lee, I. N. W.; Zhou, Y.; Benkovic, S. J.; Winograd, N. *Science* **1994**, *264*, 399.
82. Stevanovic, S.; Wiesmüller, K.-H.; Metzger, J.; Beck-Sickinger, A. G.; Jung, G. *Bioorg. Med. Chem. Lett.* **1993**, *3*, 431.
83. Wang, G. T.; Li, S.; Wideburg, N.; Krafft, G. A.; Kempf, D. J. *J. Med. Chem.* **1995**, *38*, 2995.
84. Boger, D. L.; Tarby, C. M.; Caporale, L. H.; Myers, P. L. *J. Am. Chem. Soc.* **1996**, *118*, 2109.
85. Cheng, S.; Comer, D. D.; Williams, J.-P., Myers, P. L.; Boger, D. L. *J. Am. Chem. Soc.* **1996**, *118*, 0000.

86.   Cheng, S.; Tarby, C. M.; Comer, D. D.; Williams, J. P.; Caporale, L. H.;
       Myers, P. L.; Boger, D. L. In press.
87.   Trost, B. M. *Acc. Chem. Res.* **1990**, *23*, 24 and references therein.

Chapter 10

Structurally Homogeneous and Heterogeneous Libraries

Scaffold-Based Libraries and Libraries Built Using Bifunctional Building Blocks

Viktor Krchňák, Aleksandra S. Weichsel, Dasha Cabel, and Michal Lebl

Selectide Corporation, 1580 East Hanley Boulevard, Tucson, AZ 85737

With examples of cyclic peptides and dioxopiperazines we document the strategy of building synthetic combinatorial libraries that share one common structural unit, a scaffold. These libraries we call structurally homogeneous in contrast to those that are structurally heterogeneous - those designed and built of structurally unrelated bifunctional building blocks connected by different polymer-supported chemistries. Several examples of heterogeneous libraries are shown, including N-(alkoxy acyl) amino acids, N,N'-bis(alkoxy acyl) diamino acids, N-acyl amino ethers, N-(alkoxy acyl) amino alcohols, N-alkyl amino ethers, and N-(alkoxy aryl) diamines. Polymer-supported N-acylation, etherification, esterification, reductive amination, and nucleophilic displacement have been used to synthesize these libraries.

In the middle of the last century gold was discovered at Sutter's mill in California by the American pioneer James Wilson Marshall. Gold hunters called Forty-Niners started the famous gold rush in the West that swept numerous peaceful areas and brought a lot of excitement as well as disappointment. It took more than one century before the West was in fever again. The reason was once more a vision of wealth, this time attained via combinatorial chemistry that promised to accelerate the drug discovery process enormously. And again, there was a lot of excitement as well as disappointment. Since the early times of combi-chem in San Diego (*1*) and Tucson (*2*), the gold fever, with a rush to find the mother load, a synthetic combi-chem collection from which would be mined biologically active and unique compounds that could be developed into drugs, infected pharmaceutical laboratories around the world.

1054–7487/96/0099$15.00/0

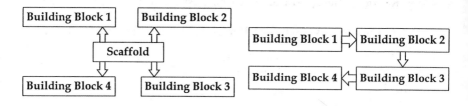

Structurally Homogenous Libraries

Structurally Heterogeneous Libraries

Figure 1. The concept of structurally homogeneous and heterogeneous libraries. Structurally homogeneous libraries are characterized by monofunctional building blocks attached to a common central unit, a scaffold. In heterogeneous libraries bifunctional building blocks are linked to each other using the same or different chemistries.

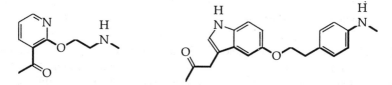

Figure 2. Structural dissimilarity of two representative structures from a library composed of aromatic hydroxy acids and amino alcohols.

In this contribution we would like to show several examples of our approach to synthetic structural diversity via combinatorial library synthesis. Instead of providing an exhaustive list of all libraries we have designed and synthesized at Selectide, we describe our overall strategy documented with examples that are characteristic for certain library types. Nevertheless, interested readers can find a more complete listing of the libraries we have made from other sources (*3-23*).

At Selectide, combinatorial chemistry is used to provide two kinds of libraries. Generic (unbiased) libraries are used to discover a novel structural motif or a feature that possesses desired biological activity, either to replace already existing drug, or to search for lead compounds for new targets, where no small molecule organic compounds with desired properties is known. Dedicated, secondary, or biased libraries serve the purpose of fine-tuning the properties of already existing lead compounds. In this article we will be dealing with the first type of libraries only, i.e. generic unbiased libraries.

The search for pharmacologically interesting compounds can be carried out among structures that have already provided successful drugs. Benzodiazepines can serve as the best example known so far (*24-26*). Alternatively, the search can focus on design and synthesis of generic libraries of diverse and previously unexplored structures to discover pharmacophores of completely novel structure. Both approaches have been applied at Selectide.

The classification of generic synthetic combinatorial libraries that we use is based on structural characteristics of individual library members. Libraries with a common central unit, a scaffold, that is present in all library members are termed structurally homogeneous. This term stresses the uniformity of all library members as far as the presence of one central scaffold unit is concerned. Libraries of another type are composed of structurally unrelated bifunctional building blocks. Such libraries do not share structure. Joints connecting these building units are the only common feature of library members. We refer to these libraries as structurally heterogeneous. Both types of libraries are depicted schematically in Fig. 1.

It is worth mentioning that peptides belong to the first category even though they are built of bifunctional building blocks, amino acids. Peptides are composed of structurally related N-alpha amino acids and accordingly share a common structural backbone, a feature which contributes to the relatively low diversity of structural space that peptides can probe. On the other hand, the use of structurally unrelated bifunctional building blocks, that is, blocks with full structural variety between the two functional groups, results in a library for which there is no common "backbone" feature. The diversity of compounds within this kind of library increases substantially. The dissimilarity of "backbone" within one such library is documented with an example of two compounds composed of structurally unrelated aromatic hydroxy acids connected to amino alcohols (Fig. 2).

Structurally homogeneous libraries

There are at least two ways to introduce the central scaffold unit into library compounds. A suitably protected scaffold can be synthesized independently and attached to the resin beads for further elaboration (e.g., cyclopentane (*8*), cyclohexane (*13*)), or the scaffold can be assembled during library synthesis (e.g., benzodiazepine (*24-26*), diketopiperazine (*27-29*), thiazolidine (*21*), pyrrolidine (*30*), benzylpiperazine (*31*)). We describe the design and synthesis of model compounds for a generic library of cyclic peptides and 2,5-disubstituted dioxopiperazines as two examples of structurally homogeneous libraries for which the scaffold is built during library synthesis.

Library of cyclic peptides.   It is no longer fashionable in pharmaceutical drug discovery research to include peptides as high priority compounds. Although, we still believe that peptides may become important drugs in some special instances. Peptides have two particularly favorable characterists. (i) The often claimed disadvantage, the tendency toward short halflife for activity of peptides, becomes an advantage in some cases (e.g., oxytocin in delivery). (ii) Peptides are composed of amino acids and as such are fully degraded along the salvage pathways to non toxic compounds.

Linear peptides are very flexible compounds; adoption of a fixed conformation occurs as an exception rather than as a rule. Cyclization within a peptide brings a constraint that has at least two consequences: (i) the peptide becomes dramatically limited in the number of conformations that it can adopt, and (ii) the peptide becomes more resistant to proteolytic cleavage and therefore it will tend to have prolonged activity in vivo. Intramolecular cyclization of a peptide is a challenging task from the chemical point of view, as documented in a number of articles dealing with polymer-supported cyclization in the last two years (*32-44*). Yet it is a task well worth the trouble, since the possibility exists that locking a peptide into a smaller subset of conformations with one of many possible methods of ring closure could produce a more active compound. Hence cyclic peptides are a very good choice of combinatorial library design (*45,46*).

In our generic cyclic peptide library we designed a lactam bridge between carboxyl and amino groups attached to the side chains of the carboxy terminal and amino terminal amino acids of a linear peptide. Our intention was to include all types and size varieties of the lactam bridge. There are four different types of lactam bridge connecting the N-terminal and the C-terminal amino acids, as shown in Fig. 3. (i) The side chains involved in the cyclization connect both alpha carbons. This represents the most common type of peptide cyclization via lactam bridge formation. Amino acids used for this type are diamino acids Lys, Orn, Dab and Dap and amino diacids Asp and Glu. The size of the lactam bridge depends on the number of methylene groups in the side chain of the amino acid involved in the cyclization. (ii) The second type of cyclic peptides connects both nitrogens. The amino acids forming this cycle are, e.g., iminodiacetic acid and N-(2-aminoethyl)glycine (A general method for this type of cyclization has been described recently (*39*)). Hybrids of the first two types represent (iii) connection of the alpha carbon to the nitrogen, and (iv) bridging the nitrogen to the alpha carbon. Each of these four types of cyclizations can have the amide bond lactam bridge in either direction; -CO-NH- or -NH-CO- (Fig. 3), and we have synthesized model compounds for both types. The head to tail cyclization belongs to the first type, since the bridge connects the alpha carbons. We excluded this cyclization, because we needed the free amino terminal amino group for the structure determination by Edman degradation (*47*) and we used the carboxy terminus for attaching the peptide to the resin-bound linker.

We intended to include both L and D amino acids in the peptide chain and still be able to determine the chirality of each amino acid. Chiral sequencing has been described recently, but we have not yet accommodated this method. We therefore used a different set of L and D amino acids. This solution might seem strange but there is logic in support of it. The dissimilarity between two peptides having amino acids differing by one methylene group is very small when compared to the dissimilarity between peptides having either an L or D amino acid in their backbone. The change in chirality does not only flip the side chain by 109 degrees, but, more importantly, it changes the conformational space that this peptide can accommodate. We believe it is legitimate to speculate that in most instances changing a peptide by adding or subtracting one methylene group from one residue's side chain will have little effect on activity. So we eliminated amino acids differing by one methylene group from the list of L-amino acids used, and replaced them with their D conformer instead. Even if this may not be true, combining the L and D amino acids in one library increases the diversity of library substantially when compared to the library composed of only L (or D) amino acids. Another rationale for doing so is the complexity of the library. If we would use all available L and D amino acids, the complexity of the library would increase enormously and there is no chance to synthesize even a few percent of all possible peptides. The final library will anyway not contain all possible structures, so limiting its complexity by eliminating the other enantiomer of the same amino acid seemed a good choice.

The sequence of a peptide can be determined by Edman degradation (*47*). The method is well developed and one bead provides enough material for automated micro-sequencing. When the amino terminal amino acid is cyclized to another residue Edman degradation proceeds. However, the hydantoin derivative formed remains bound to the solid support, since the side-chain of the N-terminal amino acid remains covalently attached. Sequencing does not reveal any amino acid in the first cycle (unless the cyclization did not proceed). However, the next cycle provides normal signal of the second amino acid. To know what N-terminal amino acid was involved in the lactam bridge, one does not recombine the library resin beads after the last amino acid coupling and finishes the synthesis (Fmoc deprotection, cyclization, side-chain deprotection) in separate pools. Then the structure of the first amino acid is given by the pool from which the bead of interest originated. In the last cycle the amino acid involved in the cyclization is cleaved and released from the bead. However, this amino acid has the amino terminal amino acid attached to its side chain. It may be in principle be possible to synthesize all of these compounds and evaluate their behavior during sequencing (retention time on HPLC). Instead, to determine the structure of the carboxy terminal amino acid we used pre-coding. A sequenceable alpha amino acid (coding for the carboxyterminal amino acid) was coupled to the coding arm prior coupling of the carboxyterminal amino acid. Edman degradation extended by one cycle revealed the code for the amino acid used in the cyclization.

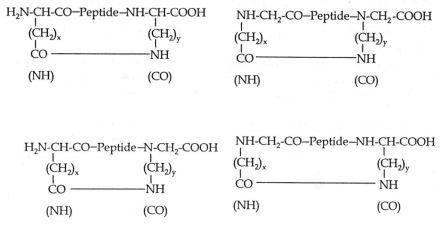

Figure 3. Four types of lactam bridge in cyclic peptides.

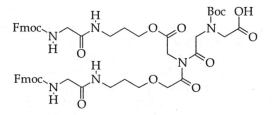

Figure 4. Structure of double cleavable linker.

Synthesis. To prove the concept, we synthesized model compounds for the generic cyclic library. The synthesis was carried out on the iminodiacetic acid (Ida) based double cleavable linker (*48*) (IdaDC, for structure see Fig. 4) that allows release of compound from a bead in two stages and therefore enables high throughput testing of released compounds in biological assyas (*49*). First compounds can be tested as mixtures and then retested with the second release as single compounds. Three arms (two screening arms and one sequencing arm) of the same peptide are built during synthesis of the library. After the two release steps for the solution phase assay, there is still one copy of the peptide attached to the resin for sequencing. The overall synthetic strategy is depicted in Fig. 5.

We used PEG grafted polystyrene/divinylbenzene resin, called TentaGel (*50*), a standard Fmoc/tBu protection strategy, and allyl ester protection of the side chain carboxyl groups involved in the cyclization. Boc-Lys(Fmoc) was coupled to the amino TentaGel resin beads, the Fmoc group was cleaved, and the coding Fmoc amino acid was coupled. The Boc group was cleaved from the α-amino group of Lys and the Fmoc protected IdaDC linker was coupled. Then the Fmoc groups from the IdaDC linker and from the coding amino acid were removed at the same time and the carboxy terminal amino acid was coupled to the amino groups on three branches (two branches on the IdaDC linker and one branch on the coding arm). The carboxy terminal amino acid was N-protected Fmoc amino diacid (Asp, Glu, and Ida), the side chain carboxyl group of Asp and Glu was protected as an allyl ester.

Synthesis was continued with formation of a linear tetrapeptide. In the last cycle the diamino acids whose alpha amino groups were protected by Boc and the side chain amino groups by Fmoc group were attached. After removing the Fmoc and the allyl ester protection groups, cyclization was performed. We tested several methods of carboxyl group activation for the cyclization: DIC/HOBt, BOP/HOBt/DIEA, HBTU/HOBt/DIEA, HOAt/DIC, and PPh₃/DEAD (Mitsunobu reaction). The first two provided the best results from the standpoint of purity and yield. In the last synthetic step all side chain protecting groups were cleaved by TFA cocktail.

The direction of the lactam bridge that was formed in these compounds is NH-CO (Fig. 5). To reverse the bridge, one can interchange the carboxy and amino terminal amino acids. The protection strategy (not actual building blocks) remain the same, Alloc for the side chain amino group and allyl ester for the side-chain carboxyl group. The synthetic scheme of compounds with the second type of bridge, i.e. CO-NH direction is shown in Fig. 6. This scheme also includes the synthesis of only two arms on a double cleavable linker. In most cases, after both releases beads still contained enough peptide for microsequencing. The only amino acid on the coding arm was the amino acid used for tagging the amino acid involved in the lactam bridge formation.

It is worthy of mention, that cyclization of a small linear peptide that does not have any N-alkylated amino acid or at least one opposite enantiomer of amino acid in its sequence is very difficult and the yield of cyclic compounds is not high. The overall purity of cyclic peptides in a library is relatively low, and when a positive compound is identified in the library, it was necessary to test whether the activity resulted from the cyclic product.

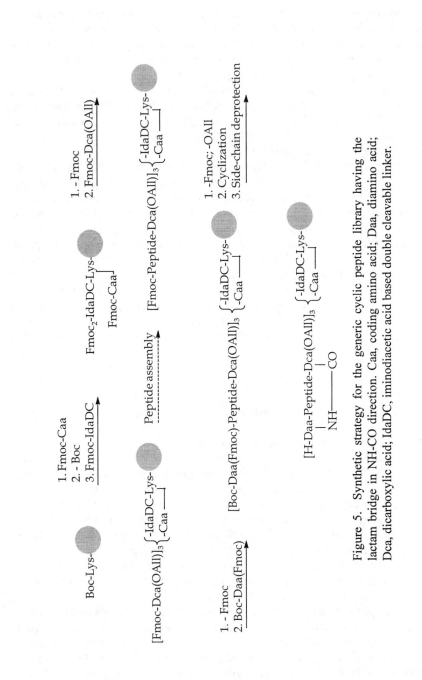

Figure 5. Synthetic strategy for the generic cyclic peptide library having the lactam bridge in NH-CO direction. Caa, coding amino acid; Daa, diamino acid; Dca, dicarboxylic acid; IdaDC, iminodiacetic acid based double cleavable linker.

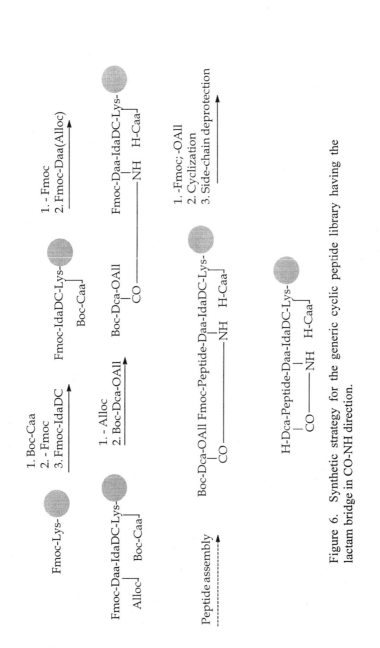

Figure 6. Synthetic strategy for the generic cyclic peptide library having the lactam bridge in CO-NH direction.

Dioxopiperazine library. Dioxopepiperazines can be considered as the smallest cyclic peptides, where the head to tail cyclization of a dipeptide forms a six-membered ring. Even if 80 N-alpha amino acids, that are currently commercially available in protected form, will be employed in this library, its complexity will still be quite limited. To increase the size of the library, we incorporated two additional randomization points: alkylation of both amino groups. A similar design has recently been published by Gordon and Steele (*28*).

The reaction scheme of the synthesis of 2,5-disubstituted dioxopiperazines on solid support is shown in Fig. 7. An fmoc amino dicarboxylic acid (Asp, Glu, Ida) was attached to the resin via its side-chain carboxyl group. The alpha carboxyl group of Asp and Glu was protected by allyl ester, Ida is symmetrical and was coupled via its anhydride. Then the amino protecting group was removed. To introduce the alkyl groups we applied reductive alkylation of polymer-supported amino group using a variety of aldehydes. We performed the condensation of aldehyde and amine in trimethylorthoformate (*51*) and we used sodium triacetoxyborohydride for reduction (*28*). Then we acylated the secondary amino group with an amino acid. We faced serious problems in driving the acylation to completion. None of the typically used procedures, such as DIC/HOBt, BOP, HBTU, HATU, and preformed symmetrical anhydrides, provided satisfactory results. We were not able to reproduce acylation of the secondary amino group as previously reported (*28*) even if we used acid fluorides prepared by a recently described procedure (*52*). Therefore we skipped the alkylation procedure to be able to acylate the first amino acid.

The synthesis continued by cleavage of the Fmoc groups from the second amino acid and the allyl ester protection from the first amino acid. The amino group was again alkylated using the same reductive alkylation procedure. The dioxopiperazine ring closed readily upon mild activation of the free carboxyl group by DIC/HOBt.

The 2,5-disubstituted dioxopiperazines do not necessarily have to be used as the sole structure in the library but they can be combined with other structural components. When a diamino acid (Lys, Orn, Dab, or Dap) is employed as the second amino acid, its side-chain amino group can be used as a starting point for building an adjoining structural unit. The amino group can be, e.g., acylated by carboxylic acids (route a in Fig. 8), the second dioxopiperazine ring can be built the same way as the first one (route b), the amino group can be alkylated by reacting with aldehydes and subsequent reduction of the Schiff base (route c), etc.

Structurally heterogeneous libraries

This type of library is built by connecting structurally unrelated bifunctional building blocks. To select a variety of bifunctional building blocks one needs a set of polymer-supported reactions to connect two functional groups. To start, we have chosen five types of organic compounds: amines, acids, aldehydes, alcohols and

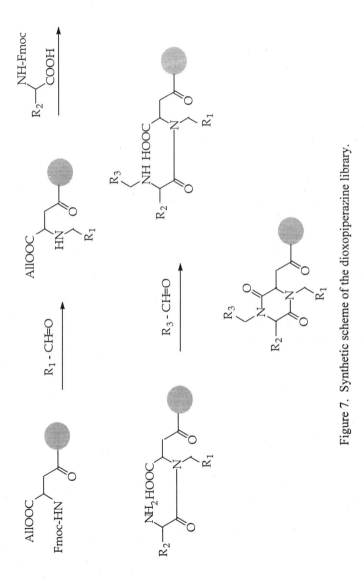

Figure 7. Synthetic scheme of the dioxopiperazine library.

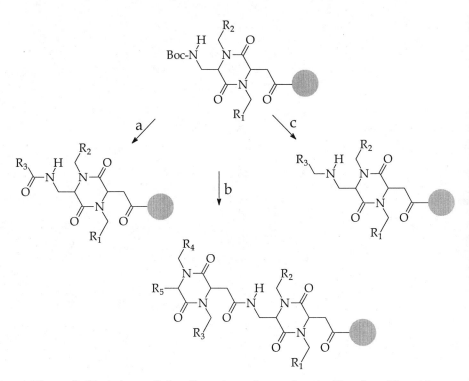

Figure 8. Variations of the dioxopiperazine moiety in libraries. The side-chain amino group is acylated (route a), alkylated (c), or used to attach the second dioxopiperazine ring (b).

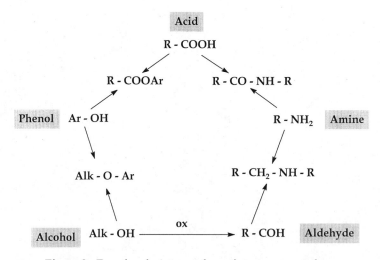

Figure 9. Functional groups and reactions to connect them.

phenols. Reactions between functional groups that we have employed in libraries are shown in Fig. 9.

The amide bond formation is an obvious choice; thanks to R.B.Merrifield the chemistry has been mastered for the last 30 years (*53*). However, to acylate the secondary amino group or an aniline type of amino group one needs to work out the conditions. For these types of amino groups the fluorides (*54*) worked best in our hands (with the exeption of alkylated N^α amino acid). For the "reversed" type of acylation, i.e., having the polymer-supported carboxyl groups, we used HBTU activation. The fluorides also provided good results.

The esterification of resin-bound hydroxyl groups by acid in solution has been used for a long time (for review see e.g. (*55*)) for attachment of the carboxyterminal amino acid to the polymer. We have found that the Mitsunobu reaction conditions can be successfully applied for this synthetic step (*56*). For the reverse esterification with polymer-bound carboxylates we found the Mitsunobu esterification to be the best procedure as far as the conversion is concerned.

Polymer-supported reductive alkylation (amino group on the resin) has been used by us and others (*10,12,28*). We found the use of trimethylorthoformate (*51*) for the formation of Schiff's base superior to the previous protocols (*10,12,28*), the condensation under these conditions is compatible with a larger variety of aldehydes. We used sodium triacetoxyborohydride for reduction of the Schiff's base (*28*).

The polymer-supported Mitsunobu etherification has been described by us (*16,17*) and others (*57*). This reaction is ideal for combinatorial synthesis, because a huge variety of alcohols are commercially available, including alcohols having a second functional group, especially an amino or a carboxyl group. Further, the yield and purity of products are good to excellent, and the reaction conditions are user-friendly.

The alcohol function on the resin can be oxidized to polymer-supported aldehydes/ketones (*17,58*). This simple and quantitative reaction serves to convert alcohols to reactive species that can be used in combinatorial chemistry in several different ways. We have applied, for example, the Wittig/Horner/Emmons reaction (*17*), recently also used by Chen et al (*58*).

Fifteen different bifunctional building units can be formed from the five functional groups listed above, as documented in Fig. 10. For library synthesis we have tested and used the bifunctional building blocks that are in italics. Each bifunctional building block was tested for its expected reactivity on simple model compounds. Only those that provided high to excellent yield and purity of product were selected for use in library synthesis. Representative examples of selected building blocks were then tested in model compounds for a particular library to prove feasibility and compatibility of use with the chemistries applied in the library.

Library design. Using the tested sets of bifunctional building blocks and polymer-supported chemical reactions for their connection one can design a variety of libraries. We limited the choices by applying the following rules for new library design: (i) Each basic generic library consisted of three randomization steps providing a complexity of

	Amine	Acid	Aldehyde	Alcohol	Phenol
Amine	*Diamines*				
Acid	*Amino acid*	*Diacid*			
Aldehyde	Amino aldehyde	*Carboxy aldehyde*	Dialdehyde		
Alcohol	*Amino alcohol*	Carboxy alcohol	Hydroxy aldehyde	Diol	
Phenol	*Amino phenol*	*Aromatic hydroxy acid*	*Phenol aldehyde*	Hydroxyalkyl phenol	Aromatic diol

Figure 10. Bifunctional building blocks containing two out of five functional groups.

125,000 compounds if 50 building blocks were used for each chemical reaction. (ii) The molecular weight of compounds present in the library was kept low, typically not exceeding 900. The average was 500. (iii) The last randomization involves either monofunctional or bifunctional building blocks, with the possibility to continue with a fourth randomization in the latter case. (iv) Linkers were used allowing the release of compounds from their beads of origin to facilitate the use of assays that require compounds to be tested in solution. (v) Library beads were not combined after the last randomization to simplify post-screening structure determination. (vi) Mass spectroscopy was used to determine the molecular weight, which information contributed to structure elucidation.

Example of libraries. The general structures of six libraries are shown in Fig. 11. The first two libraries share chemistry and building blocks: amino acids, aromatic hydroxy acids, and alcohols. The synthesis of the library of N-(alkoxy acyl) amino acids started with an amino acid randomization, then the aromatic hydroxy acids were attached to the polymer-supported amino groups in the second synthetic cycle. The phenolic hydroxyl groups were alkylated under the Mitsunobu reaction conditions in the last cycle. The library of N,N'-bis(alkoxy acyl) diamino acids employed diamino acids (Lys, Orn, Dab, Dap) instead of amino acids. After the acylation of the alpha amino group and alkylation of aromatic hydroxyls, we deprotected the side-chain amino group of diamino acid (protected by Alloc group) and repeated the sequence of two reactions on the second amino group. Both libraries have been synthesized on OH-TentaGel and on IdaDC derivatized NH_2-TentaGel.

The next two libraries have been designed using aromatic hydroxy acids and amino alcohols as the bifunctional building blocks. The library of N-acyl amino ethers started with randomization of aromatic hydroxy acids, followed by alkylation of resin-supported phenols by N-Fmoc protected amino alcohols. The last cycle of the synthesis involved acylation of the deprotected amino groups by a set of carboxylic acids. The synthesis of the library of N-alkyl amino ethers differed from the previous synthesis only in the last cycle, where the amino groups were alkylated by reductive alkylation. IdaDC derivatized resin was used for both libraries.

The N-(alkoxy acyl) amino alcohols have been synthesized on carboxy derivatized resin support. The synthesis started with an esterification of polymer-supported carboxylates by N-protected (Fmoc) amino alcohols. We have tested several procedures for this "reverse" polymer-supported esterification, the only procedure that provided good conversion to esters (greater than 80 %) was the Mitsunobu PPh_3/DEAD method. After removing the Fmoc protecting group, the amino group was acylated by a set of aromatic hydroxy acids. The phenolic hydroxyl groups were alkylated by alcohols in the last synthetic cycle. Since the compounds were attached to the beads via an ester bond, we used diluted sodium hydroxide or ammonia gas to release the compounds from the beads into for screening. This library provided compounds that all shared one hydroxyl group left behind by the linker.

The N-(alkoxy aryl) diamines are included to represent non-amide bond structure. The library was synthesized on trityl linker and started with reaction of diamines with the chlorotrityl resin. In the presence of a high excess of diamines

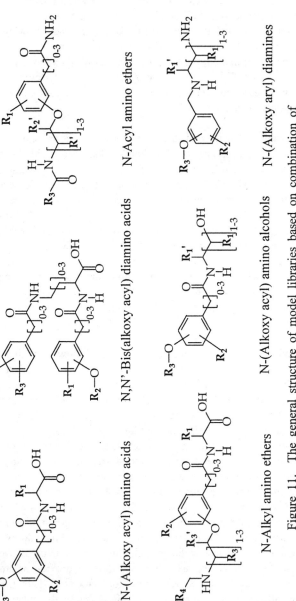

Figure 11. The general structure of model libraries based on combination of bifunctional building blocks.

predominantly monoalkylated species were formed. The amino group was then alkylated by aromatic hydroxy aldehydes by reductive alkylation. In the last cycle the phenols were alkylated by alcohols. The compounds were cleaved from the linker by trifluoroacetic acid, either in solution or by TFA vapors. Hydrogen chloride gas was also tested as an alternative to TFA cleavage. An analogous library can be synthesized with N-protected amino alcohols or amino acids are used in the first synthetic cycle instead of the diamines.

Conclusions

We have documented with several examples the design and synthesis of two types of synthetic combinatorial libraries. Structurally homogenous libraries, characterized by the presence of one common structural feature in all library compounds, a scaffold, were represented by cyclic peptides and dioxopiperazines. In structurally heterogeneous libraries different organic reactions (acylation, etherification, reductive amination, nucleophilic displacement) were applied to connect bifunctional types of building blocks (aromatic hydroxy acids, aromatic hydroxy aldehydes, amino alcohols, diamines, diacids, and amino acids).

References

1. Houghten, R. A.; Pinilla, C.; Blondelle, S. E.; Appel, J. R.; Dooley, C. T.; Cuervo, J. H. *Nature* **1991**, *354*, 84-86.
2. Lam, K. S.; Salmon, S. E.; Hersh, E. M.; Hruby, V. J.; Kazmierski, W. M.; Knapp, R. J. *Nature* **1991**, *354*, 82-84.
3. Lam, K. S.; Zhao, Z. G.; Wade, S.; Krchnak, V.; Lebl, M. *Drug Develop. Res* **1994**, *33*, 157-160.
4. Lam, K. S.; Lebl, M.; Krchnak, V.; Lake, D. F.; Smith, J.; Wade, S.; Ferguson, R.; Ackerman-Berrier, M.; Wertman, K.In *Peptides. Proceedings of the Thirteenth American Peptide Symposium*; Hodges, R. S., Smith, J. A. Eds.; ESCOM: Leiden, 1994; pp 1003-1004.
5. Lebl, M.; Krchnak, V.; Salmon, S. E.; Lam, K. S. *Methods. A companion to Methods in Enzymology* **1994**, *6*, 381-387.
6. Lebl, M.; Krchnak, V.; Safar, P.; Stierandova, A.; Sepetov, N. F.; Kocis, P.; Lam, K. S. *Techniques in Protein Chemistry* **1994**, *5*, 541-548.
7. Lebl, M.; Krchnak, V.; Sepetov, N. F.; Kocis, P.; Patek, M.; Flegelova, Z.; Ferguson, R.; Lam, K. S. *J. Protein Chem.* **1994**, *13*, 484-486.
8. Patek, M.; Drake, B.; Lebl, M. *Tetrahedron Lett.* **1994**, *35*, 9169-9172.
9. Stankova, M.; Wade, S.; Lam, K. S.; Lebl, M. *Peptide Res* **1994**, *7*, 292-298.
10. Stankova, M.; Issakova, O.; Sepetov, N. F.; Krchnak, V.; Lam, K. S.; Lebl, M. *Drug Develop. Res* **1994**, *33*, 146-156.
11. Vagner, J.; Krchnak, V.; Sepetov, N. F.; Strop, P.; Lam, K. S.; Barany, G.; Lebl, M.In *Innovation & Perspectives in Solid Phase Synthesis*; Epton, R. Ed.; Mayflower Worldwide Limited: Birmingham, 1994; pp 347-352.

12. Flegelova, Z.; Krchnak, V.; Sepetov, N. F.; Stankova, M.; Issakova, O.; Cabel, D.; Lam, K. S.; Lebl, M.In *The Peptides 1994*; Maia, H. L. S. Ed.; ESCOM: Leiden, 1995; pp 469-470.

13. Kocis, P.; Issakova, O.; Sepetov, N. F.; Lebl, M. *Tetrahedron Lett.* **1995**, *36*, 6623-6626.

14. Krchnak, V.; Sepetov, N. F.; Kocis, P.; Patek, M.; Lam, K. S.; Lebl, M.In *Combinatorial libraries. Synthesis, screening and application potential*; Cortese, R. Ed.; Walter de Gruyter: Berlin, 1996; pp 27-52.

15. Krchnak, V.; Weichsel, A. S.; Cabel, D.; Lebl, M. *Peptide Res* **1995**, *8*, 198-204.

16. Krchnak, V.; Flegelova, Z.; Weichsel, A.; Lebl, M. *Tetrahedron Lett.* **1995**, *36*, 6193-6196.

17. Krchnak, V.; Vagner, J.; Flegelova, Z.; Weichsel, A. S.; Barany, G.; Lebl, M.In *Peptides. Proceedings of the Fourteenth American Peptide Symposium*; Kaumaya, P. T. P. Ed.; ESCOM: Leiden, 1995; pp in press

18. Lam, K. S.; Wade, S.; Abdul-Latif, F.; Lebl, M. *J Immunol Methods* **1995**, *180*, 219-223.

19. Lebl, M.; Krchnak, V.; Sepetov, N. F.; Seligmann, B.; Strop, P.; Felder, S.; Lam, K. S. *Biopolymers* **1995**, *37*, 177-198.

20. Madden, D.; Krchnak, V.; Lebl, M. *Perspectives in Drug Discovery and Design* **1995**, *2*, 269-285.

21. Patek, M.; Drake, B.; Lebl, M. *Tetrahedron Lett.* **1995**, *36*, 2227-2230.

22. Seligmann, B.; Abdul-Latif, F.; Al-Obeidi, F.; Flegelova, Z.; Issakova, O.; Kocis, P.; Krchnak, V.; Lam, K.; Lebl, M.; Ostrem, J.; Safar, P.; Sepetov, N.; Stierandova, A.; Strop, P.; Wildgoose, P. *Eur. J Med. Chem.* **1995**, *30*, S319-S335.

23. Sepetov, N. F.; Krchnak, V.; Stankova, M.; Wade, S.; Lam, K. S.; Lebl, M. *Proc. Natl. Acad. Sci. USA* **1995**, *92*, 5426-5430.

24. Bunin, B. A.; Plunkett, M. J.; Ellman, J. A. *Proc. Natl. Acad. Sci. USA* **1994**, *91*, 4708-4712.

25. DeWitt, S. H.; Schroeder, M. C.; Stankovic, C. J.; Strode, J. E.; Czarnik, A. W. *Drug Develop. Res* **1994**, *33*, 116-124.

26. Plunkett, M. J.; Ellman, J. A. *J Am. Chem. Soc.* **1995**, *117*, 3306-3307.

27. Safar, P.; Stierandova, A.; Lebl, M.In *The Peptides 1994*; Maia, H. L. S. Ed.; ESCOM: Leiden, 1995; pp 471-472.

28. Gordon, D. W.; Steele, J. *Bioorg. Med. Chem.* **1995**, *5*, 47-50.

29. Terrett, N. K.; Bojanic, D.; Brown, D.; Bungay, P. J.; Gardner, M.; Gordon, D. W.; Mayers, C. J.; Steele, J. *Bioorg. Med. Chem. Lett.* **1995**, *5*, 917-922.

30. Murphy, M. M.; Schullek, J. R.; Gordon, E. M.; Gallop, M. A. *J Am. Chem. Soc.* **1995**, *117*, 7029-7030.

31. Dankwardt, S. M.; Newman, S. R.; Krstenansky, J. L. *Tetrahedron Lett.* **1995**, *36*, 4923-4926.

32. Osapay, G.; Bouvier, M.; Taylor, J. W. *Techniques in Protein Chemistry* **1991**, *2*, 221-231.

33. Rovero, P.; Quartara, L.; Fabbri, G. *Tetrahedron Lett.* **1991**, *32*, 2639-2642.

34. Ehrlich, A.; Rothemund, S.; Brudel, M.; Beyermann, M.; Carpino, L. A.; Bienert, M. *Tetrahedron Lett.* **1993**, *34*, 4781-4784.

35. Kapurniotu, A.; Taylor, J. W. *Tetrahedron Lett.* **1993**, *34*, 7031-7034.
36. Marlowe, C. K. *Bioorg. Med. Chem. Lett.* **1993**, *3*, 437-440.
37. Mcmurray, J. S.; Lewis, C. A. *Tetrahedron Lett.* **1993**, *34*, 8059-8062.
38. Alsina, J.; Rabanal, F.; Giralt, E.; Albericio, F. *Tetrahedron Lett.* **1994**, *35*, 9633-9636.
39. Kaljuste, K.; Unden, A. *Int. J Pept. Protein Res* **1994**, *43*, 505-511.
40. Kates, S. A.; Sole, N. A.; Johnson, C. R.; Hudson, D.; Barany, G.; Albericio, F. *Tetrahedron Lett.* **1994**, 4
41. Mcmurray, J. S.; Lewis, C. A.; Obeyesekere, N. U. *Peptide Res* **1994**, *7*, 195-206.
42. Richter, L. S.; Tom, J. Y. K.; Burnier, J. P. *Tetrahedron Lett.* **1994**, *35*, 5547-5550.
43. Zhao, Z. C.; Felix, A. M. *Peptide Res* **1994**, *7*, 218-223.
44. Kania, R. C.; Zuckermann, R. N.; Marlowe, C. K. *J Am. Chem. Soc.* **1994**, *116*, 8835-8836.
45. Darlak, K.; Romanovskis, P.; Spatola, A. F.In *Peptides,Proc.13th APS*; Hodges, R. S., Smith, J. A. Eds.; ESCOM,Leiden: 1994; pp 981-983.
46. Eichler, J.; Lucka, A. W.; Houghten, R. A. *Peptide Res* **1994**, *7*, 300-307.
47. Edman, P. *Acta Chem. Scand.* **1950**, *4*, 283-293.
48. Kocis, P.; Krchnak, V.; Lebl, M. *Tetrahedron Lett.* **1993**, *34*, 7251-7252.
49. Salmon, S. E.; Lam, K. S.; Lebl, M.; Kandola, A.; Khattri, P. S.; Wade, S.; Patek, M.; Kocis, P.; Krchnak, V.; Thorpe, D.; Felder, S. *Proc. Natl. Acad. Sci. USA* **1993**, *90*, 11708-11712.
50. Bayer, E.; Rapp, W.In *Poly(Ethylene Glycol) Chemistry: Biotechnical and Biomedical Applications*; Harris, J. M. Ed.; Plenum Press: New York, 1992; pp 325-345.
51. Look, G. C.; Murphy, M. M.; Campbell, D. A.; Gallop, M. A. *Tetrahedron Lett.* **1995**, *36*, 2937-2940.
52. Tamamura, H.; Otaka, A.; Nakamura, J.; Okubo, K.; Koide, T.; Ikeda, K.; Ibuka, T.; Fujii, N. *Int. J Pept. Protein Res* **1995**, *45*, 312-319.
53. Merrifield, R. B. *J Am. Chem. Soc.* **1963**, *85*, 2149-2154.
54. Carpino, L. A.; Sadataalaee, D.; Chao, H. G.; Deselms, R. H. *J Am. Chem. Soc.* **1990**, *112*, 9651-9652.
55. Fields, G. B.; Noble, R. L. *Int. J Pept. Protein Res* **1990**, *35*, 161-214.
56. Krchnak, V.; Cabel, D.; Weichsel, A. S.; Flegelova, Z. *Lett. Peptide Sci.* **1995**, *2*, 277-282.
57. Rano, T. A.; Chapman, K. T. *Tetrahedron Lett.* **1995**, *36*, 3789-3792.
58. Chen, C. X.; Randall, L. A. A.; Miller, R. B.; Jones, A. D.; Kurth, M. J. *J Am. Chem. Soc.* **1994**, *116*, 2661-2662.

Chapter 11

Liquid-Phase Combinatorial Synthesis

Dennis J. Gravert and Kim D. Janda[1]

Departments of Molecular Biology and Chemistry, Scripps Research Institute, 10666 North Torrey Pines Road, La Jolla, CA 92037

Liquid-phase combinatorial synthesis (LPCS) has been developed as an alternative methodology for the construction of small molecule libraries wherein the use of a *soluble* polymer support provides both the advantages of solid-phase synthesis and the benefits of classical solution-phase organic chemistry. The utility of LPCS was demonstrated by the synthesis of three different libraries containing peptide, peptidomimetic, or non-peptide molecules. The peptide-based library was synthesized and screened through a recursive deconvolution strategy, and several members were found to bind an anti-β-endorphin monoclonal antibody. The peptidomimetic library was successfully constructed by LPCS to yield a collection of oligomeric aza-amino acids or "azatides". These biopolymer mimics may serve to explore peptide structure as well as to provide a source of potential drug leads. The non-peptidyl library consists of sulfonamides, a class of compounds of known clinical bactericidal efficacy. The results indicate that LPCS should have general utility to construct small molecule libraries of high chemical diversity. Furthermore, LPCS can provide multimilligram quantites of each library member which should allow for multiple high-throughput screening assays.

Pharmaceutical research and development has undergone a widespread change in the process of obtaining new drug candidates. The traditional, arduous task of synthesizing and assaying drug analogs one molecule at a time has given way to a new process of drug discovery that employs combinatorial chemistry. In the search for new leads, the pharmaceutical industry has utilized automated high-throughput methods that allow for the parallel screening of the numerous chemical entities provided by this new field of chemistry.

Initially, synthetic peptides and nucleic acid libraries were constructed as procedures have been found to synthesize these molecules combinatorially (*1*). Yet, these libraries are not only limited in chemical diversity, but their library members suffer from less than optimal pharmacokinetic parameters required for oral delivery of drugs. For pharmaceutical development, it is imperative to increase diversity using "small molecule" libraries and thus increase the potential discovery of bioavailable

---

[1]Corresponding author

1054–7487/96/0118$15.00/0

organic compounds.   To supply the compounds for drug screening, new methodologies are needed to build these small molecule libraries.   In addition to designing methods that incorporate new molecular entities in a combinatorial format, success in these new synthetic schemes depends on achieving high-yielding reactions. Furthermore, successful implementation of combinatorial syntheses require simple product isolation, ease of the "portioning-mixing" technique (2), and ability to drive a reaction to completion.

To meet these goals, polymer-supported synthesis has become a popular strategy for the construction of non-oligopeptide/nucleotide libraries (3-6). With few exceptions, insoluble polymers supports have been used to achieve product purification after each synthetic step through simple filtration and rinsing. Although highly successful, solid-phase synthesis still has several shortcomings. The nature of the heterogeneous reaction conditions lead to nonlinear kinetic behavior, unequal distribution and/or access to the chemical reaction, solvation problems, and pure synthetic problems associated with solid phase synthesis. These limitations have prompted us to seek out alternative methodologies from a combinatorial point of view. We have discovered that the use of a polymer support which is soluble during the course of a reaction yet may be later precipitated and filtered allows for the development of a liquid phase method (7). This methodology, termed liquid-phase combinatorial synthesis (LPCS), in essence avoids the difficulties of solid-phase synthesis while preserving its positive aspects.

Liquid-Phase Combinatorial Synthesis (LPCS)

The success of the liquid-phase method results from the use of polyethylene glycol monomethyl ether (MeO-PEG) as the polymeric support. Employed as a protecting group, this soluble linear homopolymer has been productive in peptide, oligonucleotide, and oligosaccharide synthesis (8-10). MeO-PEG solubility in a wide variety of aqueous and organic solvents not only enables homogeneous reactions under numerous reaction conditions, but these solubility properties permit individual reactions steps to be monitored by infrared spectroscopy and/or proton and carbon-13 nuclear magnetic resonance spectroscopy.   Furthermore, this mono-functional polymer exhibits a helical structure that possesses a strong propensity to crystallize; thus, as long as the polymer remains unaltered during library synthesis then purification by crystallization can be utilized at each reaction step.

Liquid phase synthesis is amenable to combinatorial techniques; in fact, this solution-phase method permits all manipulations, including portion-mixing or split synthesis, to be performed under homogeneous conditions. The integration of liquid phase synthesis into a combinatorial format results in the method we term "liquid phase combinatorial synthesis" (LPCS). The validity of this method is demonstrated first by the synthesis of a peptide library from which ligands were identified that bind an acceptor molecule. The general utility of this method is revealed through synthesis of two non-peptide libraries. Thus by providing small molecule libraries of "α-aza-amino acids" and sulfonamides, LPCS extends the types of chemistry and the classes of compounds accessible for the screening of drug candidates.

Peptide Library through LPCS

Synthesis of a combinatorial peptide library was achieved using LPCS and a recursive deconvolution methodology (7, 11). In this protocol, partially synthesized libraries are obtained by saving and cataloging aliquots during the synthesis of the combinatorial library. These partial libraries are later utilized in the screening process to identify novel library members with targeted properties.

To demonstrate the utility of LPCS, a library of 1024 pentapeptides composed of four amino acids (Tyr, Gly, Phe, Leu) with five partial libraries were synthesized through this methodology. Using an affinity assay, this library was screened for high affinity ligands to a monoclonal antibody that binds the β-endorphin sequence Tyr-Gly-Gly-Phe-Leu with great affinity ($K_d = 7.1$ nM) (12).

Library Synthesis. A variation of the split synthesis approach was employed to allow recursive deconvolution of the combinatorial library to identify peptides with high binding properties ($7, 11$). Figure 1 lists the steps required to produce the pentapeptide library containing an alphabet of four different amino acids. Synthesis began by coupling each member (Tyr, Gly, Phe, Leu) of the alphabet to the polymeric support (MeO-PEG) in segregated reaction vials to establish four channels of synthesis. After completion of N,N'-dicyclohexylcarbodiimide mediated coupling in methylene chloride, unreacted hydroxyl end groups of MeO-PEG were end-capped using phenyl isocyanate and catalytic amounts of dibutyltinlaurate. Based on the absorbance of the phenyl carbamate derivative, the amino acid coupling efficiency was determined to be >99%. After removal of dicyclohexylurea by filtration, reaction solutions were concentrated and diluted with diethyl ether to induce precipitation/crystallization of the MeO-PEG-amino acid conjugate. Filtering and washing with ether or cold ethanol completed purification; however, if required, simple recrystallization of the MeO-PEG-coupled product removed more polar contaminates. In fact, crystallization avoids the possibility of inclusions which may occur with gelatinous precipitates, and thus excess reagents may be removed quantitatively. After dissolving in methylene chloride, a portion of the polymer solution from each reaction vial was removed, set aside, and labeled as partial library p(1) consisting of the four members MeO-PEG-Naa (where Naa is Tyr, Gly, Phe, or Leu). The remaining solutions of each reaction vial was then combined, mixed, and divided among the reaction vials. Normal peptide chemistry using *tert*-butyloxycarbonyl (Boc) protected amino acids continued library synthesis (7). After each coupling step, portions of the purified PEG-bound peptides were again sampled to generate a sublibrary labeled as p(n) where n corresponds to the particular step in the synthetic scheme. This process of coupling, saving/cataloging, and randomizing was repeated until the desired pentapeptide library was obtained.

Library Screening. The peptide library was screened for ligands to an anti-β-endorphin antibody, and identification of high affinity ligands was accomplished through recursive deconvolution as outlined in Figure 2. At the completion of synthesis, the library remained divided among the reaction vials such that each vial contained a pool of peptides whose terminal residue was known. Each pool was then analyzed using a competitive ELISA assay. The natural antibody ligand (Tyr-Gly-Gly-Phe-Leu) was attached to bovine serum albumin (BSA), and this conjugate affixed to ELISA plates (7). Analysis of each pool in solution competition experiments yielded IC_{50} values.

In the first round of screening, only one pool of the p(5) sublibrary yielded detectable binding. Because the pool MeO-PEG-Naa-Naa-Naa-Naa-Tyr gave an IC_{50} value of 51 μM, tyrosine was identified as the essential terminal residue and was coupled to the four pools of the p(4) sublibrary. Screening these four new pools not only provided an enrichment step, but identified the penultimate residue. This process of coupling and screening was then repeated until the full identities of the active peptides were found. Screening the final sublibrary identified the native epitope and several other potent binders (7).

The direct assay of PEG-peptide conjugates combined with the recursive deconvolution strategy resulted in the discovery of ligands that inhibited the binding of leucine enkephalin to anti-β-endorphin monoclonal antibody 3E7. The diverse solubilizing power of MeO-PEG provided a direct method for screening peptide libraries in a homogeneous assay. If desired, however, peptides may be cleaved from

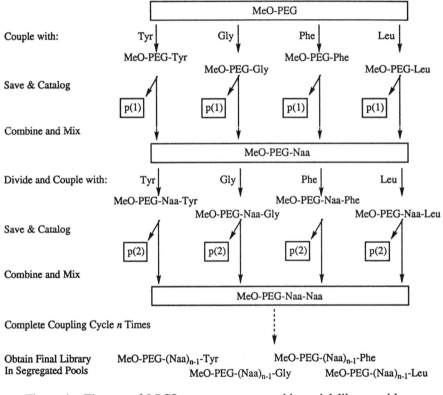

Figure 1. The use of LPCS to construct a combinatorial library with an alphabet of four amino acids (Naa = randomized position). After each nth coupling step, a portion of each pool is saved and catalogued to form a partial sublibrary $p(n)$.

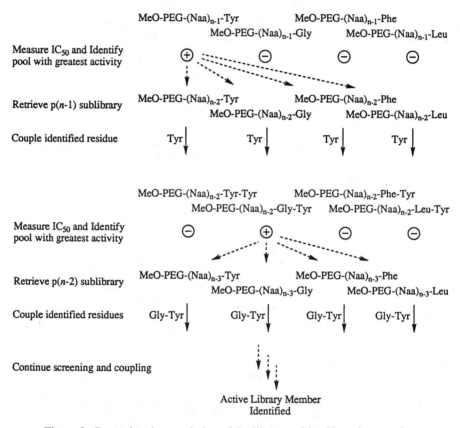

Figure 2. Recursive deconvolution of the library to identify active members.

the polymer support to permit the assay of "deprotected" libraries of ligands. In fact, the detected binding affinities were found to be similar between the p(5) sublibraries of MeO-PEG-peptide conjugates (51 μM) and free peptides (46 μM) (7).

Peptidomimetic Library of "α-Aza-Amino Acids" or "Azatides"

Liquid phase combinatorial synthesis has been applied to the construction of peptidomimetics consisting of oligomers of "α-aza-amino acids" or "azatides" (Han, H.; Janda, K. D., *J. Am. Chem. Soc.*, in press). Interest in these azatides and the study of peptidomimetics has been driven by the pursuit of bioactive molecules with improved pharmacokinetic properties. Modifications of the peptide structure, most commonly inversion of stereochemistry to yield D-amino acids (*13*), have yielded compounds with improved biological potencies, altered conformational properties (*14*), and increase resistance to enzymatic degradation (*15*).

Azapeptides are peptidomimetics that have the α-carbon of one or more residue replaced by a nitrogen atom. The resulting structure lacks the asymmetry of the normal amino acid, and the structure may be considered intermediate in conformation between D- and L-amino acids (*13, 16*). Such azapeptides have shown resistance to enzymatic cleavage and selective inhibition of cysteine (*17*) and serine proteases (*18-22*). The complete substitution of all α-carbons with nitrogen in a peptide results in a new oligomeric species which we haved termed an "azatide". In addition to contributing to the study of peptidomimetics, azatides may serve as a source of new libraries to screen for leads for pharmaceutical development. Therefore, the synthesis of α-aza-amino acids and their coupling to form azatides using LPCS was undertaken. This successful achievement demonstrated that oligoazatides can be rapidly assembled on a homogeneous polymeric support.

Synthesis of Azatide Monomers. Initial efforts focused on the development of general synthetic procedures to obtain a "new alphabet" of α-aza-amino acid monomers. In the design of Boc-protected azatide monomers, the Boc group can be placed on either nitrogen of an alkylhydrazine. Attachment of *tert*-butyloxycarbonyl to the nitrogen substituted with an alkyl group would yield an "C-protected" azatide monomer; similarly, Boc protection of the unsubstituted nitrogen generates the "N-protected" azatide monomer. Because the coupling efficiency of each isomer was unknown, synthetic routes to both of these azatide monomers were required and were successfully found (Scheme 1). The "C-protected" azatide monomer was obtained via Boc-protection of an alkylated hydrazine. "N-protection" was achieved through reduction of Boc-protected hydrazones derived from reaction of Boc-carbazate with a carbonyl. A diverse selection of alkyl substituents enables the generation of a unique alphabet of α-aza-amino acids.

Coupling of Azatide Monomers. With several azatide monomers in hand, efforts were first made to optimize the coupling of aza-amino acids under homogeneous reaction conditions in the absence of a polymeric support. As Boc-alkylhydrazines are poorer nucleophiles than simple amines or amino acids, a highly activated carbonyl synthon was required to permit efficient coupling and oligomer synthesis. Furthermore, the coupling had to be controllable to minimize symmetrical dimerization. The poor reaction yields, prolonged reaction times, and/or side reactions of *p*-nitrophenol chloroformate, carbonyldiimidazole, bis-(2,4-dinitrophenyl) carbonate, and trichloromethyl chloroformate led to use of bis-pentafluorophenyl carbonate as the activation agent. This readily prepared reagent

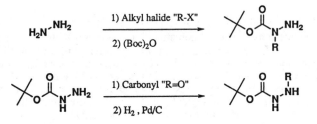

Scheme 1. Synthesis of "C-protected" (top) and "N-protected" (bottom) azatide monomers.

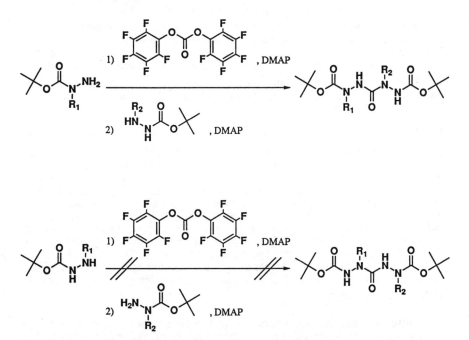

Scheme 2. Synthesis of diazatides through "N-activated" (top) and "C-activated" (bottom) carbamates. The "C-activation" route was unsuccessful.

from phosgene and pentafluorophenol is a highly crystalline compound that once formed is easy to handle.

The feasibility of general azatide coupling was demonstrated by bis-pentafluorophenyl carbonate-mediated coupling. Whereas "N-activation" led to azatides in good yield, "C-activation" was unsuccessful (Scheme 2). Under "N-activation" coupling conditions, both aza-glycine-aza-glycine (Gly^{a}Gly^a, where superscript (a) represents an α-aza-amino acid) and the more sterically demanding aza-valine-aza-valine (Val^{a}Val^a) was synthesized in less than an hour. After establishing the coupling chemistry, research was applied towards the development of a method to construct well-defined azatides using a polymeric support.

Azatides by LPCS. The design and synthesis of a linker was necessary to permit a liquid phase method of azatide synthesis in a combinatorial format. A *p*-substituted benzyl ester spacer unit was made to accommodate directional synthesis of MeO-PEG-azatides and to be compatible with Boc-chemistry. This linker was stable against acidolysis and allowed release of the oligoazatide by catalytic hydrogenation. Thus, it was possible to construct the azatide pentamer Y^{a}G^{a}G^{a}F^{a}L^a in 56.7% overall yield. Synthesis on MeO-PEG allowed the reaction progress to be monitored by proton nuclear magnetic resonance spectroscopy and by the Kaiser ninhydrin test (*23*). Furthermore, the linker strategy allowed for the deprotection of tyrosine and the release of Boc-protected azatide in a single step using catalytic hydrogenation. Treatment with trifluoroacetic acid yielded the final deprotected azatide which was analyzed by tandem mass spectroscopy for sequence determination. Mass spectral analysis based on predicted fragmentation patterns confirmed the sequence Y^{a}G^{a}G^{a}F^{a}L^a.

Non-Peptidyl Library of Sulfonamides

The LPCS method should permit the synthesis of any compound as long as the chemistry does not interact with or adversely affect MeO-PEG properties. To verify the utility of LPCS to generate non-peptidyl compounds, a library of sulfonamides has been constructed by parallel synthesis (*7*). These molecules have provided an inexpensive treatment for bacterial infections in the past. However, bacterial resistance, a narrow antibacterial spectrum, and side effects have limited their clinical usage. Actually, the observed side effects have led to studies that show promise of developing sulfonamides into antitumor agents (*24*), endothelin antagonists (*25*), and anti-arrhythmic agents (*26*). The screening of combinatorial libraries of sulfonamides may therefore led to the discovery of new pharmaceutical agents.

A new route for the synthesis of sulfonamides was developed to allow attachment of the arylsulfonyl moiety to MeO-PEG and to provide for molecular diversity (Scheme 3). Attachment to the polymeric support was formed quantitatively by reaction of 4-(chlorosulfonyl)phenyl isocyanate with MeO-PEG in the presence of a catalytic amount of dibutyltinlaurate. Impressively, coupling occurred without any competing participation of the sulfonyl chloride moiety. Splitting the MeO-PEG-arylsulfonyl chloride among six reaction vials containing different amines resulted in a library of MeO-PEG-protected sulfonamides. Hydrolytic cleavage of the urethane linkage under basic conditions yielded the final library of six members in analytically pure form in overall yields of 95-97% (*7*). It should be noted that the overall success of sulfonamide synthesis was dependent on the pK_a of the nucleophilic amine. By extending reaction times and using higher temperatures, a broad pK_a range of amines was successfully incorporated. Fortunately, the homogeneous support permitted the monitoring of reaction progress by proton nuclear magnetic resonance spectroscopy and allowed the reaction sequence to occur cleanly to produce multimilligram quantities of each sulfonamide.

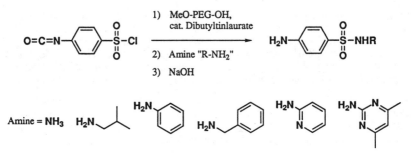

Scheme 3. The use of LPCS to construct a combinatorial library of sulfonamides.

Conclusions

The methods of LPCS and recursive deconvolution have been successfully developed to simplify and speed the synthesis and screening of small molecule combinatorial libraries for drug lead discovery. LPCS combines the advantages of solid phase synthesis with the benefits of classical organic synthesis in solution. Findings thus far demonstrate that the reaction scope of this protocol should be general and dictate that LPCS has value in the synthesis of multimilligram quantites of library members for multiple, high throughput screening assays. Furthermore, LPCS shows promise as a method to produce new libraries of highly diverse and complex molecules in a combinatorial format.

Acknowledgments

This work was supported in part by The Scripps Research Institute, the R. W. Johnson Pharmaceutical Research Institute, and the Alfred P. Sloan Foundation.

Literature Cited

1.   Janda, K. D. *Proc. Natl. Acad. Sci. USA* **1994**, *91*, 10779.
2.   Furka, A.; Sebestyen, M.; Dibo, G. *Highlights of Modern Biochemisty: Proceedings of the 14th International Congress of Biochemistry;* VSP, Ultrecht, The Netherlands, 1988; Vol. 5, p. 47.
3.   Bunin, B. A.; Ellman, J. A. *J. Am. Chem. Soc.* **1992**, *114*, 10997.
4.   Hobbs Dewitt, S.; Kiely, J. S.; Stankovic, C. J.; Schroeder, M. C.; Reynolds Cody, D. M. *Proc. Natl. Acad. Sci. USA* **1993**, *90*, 6909.
5.   Chen, C.; Ahlberg, L. A.; Miller, R. B.; Jones, A. D.; Kurth, M. J. *J. Am. Chem. Soc.* **1994**, *116*, 2661.
6.   Backes, J. B.; Ellman, J. A. *J. Am. Chem. Soc.* **1994**, *116*, 11171.
7.   Han, H.; Wolfe, M. M.; Brenner, S.; Janda, K. D. *Proc. Natl. Acad. Sci. USA* **1995**, *92*, 6419.
8.   Bayer, E.; Mutter, M. *Nature (London)* **1972**, *237*, 512.
9.   Bonora, G. M.; Scremin, C. L.; Colonna, F. P.; Garbesi, A. *Nucleic Acids Res.* **1990**, *18*, 3155.

10. Douglas, S. P.; Whitfield, D. M.; Krepinsky, J. J. *J. Am. Chem. Soc.* **1991**, *113*, 5095.
11. Erb, E.; Janda, K. D.; Brenner, S. *Proc. Natl. Acad. Sci. USA* **1994**, *91*, 11422.
12. Cwirla, S. E.; Peters, E. A.; Barrett, R. W.; Dower, W. J. *Proc. Natl. Acad. Sci. USA* **1990**, *87*, 6378.
13. Spatola, A. F. In *Chemistry and Biochemistry of Amino Acids, Peptides, and Proteins*; Weinstein, B., Ed.; Mercel Dekker: New York, NY, 1983; pp. 267-357.
14. Mosberg, H. I.; Hurst, R.; Hruby, V. J.; Gee, K.; Yamamura, H. I.; Galligan, J. J.; Burks, T. F. *Proc. Natl. Acad. Sci. USA* **1983**, *80*, 5871.
15. Dooley, C. T.; Chung, N. N.; Wilkes, B. C.; Schiller, P. W.; Bidlack, J. M.; Pasternak, G. W.; Houghton, R. N. *Science* **1994**, *266*, 2019.
16. Aubry, A.; Marraud, M. *Biopolymers* **1989**, *28*, 109.
17. Magrath, J.; Abeles, R. H. *J. Med. Chem.* **1992**, *35*, 4279.
18. Elmore, D. T.; Smyth, J. J. *Biochem. J.* **1968**, *107*, 103.
19. Barker, S. A.; Gray, C. J.; Ireson, J. C.; Parker, R. C. *Biochem. J.* **1968**, *107*, 103.
20. Gray, C. J.; Al-Dulaimi, K.; Khoujah, A. M.; Parker, R. C. *Tetrahedron* **1977**, *33*, 837.
21. Gupton, B. F.; Carroll, D. L.; Tuhy, P. M.; Kam, C.-M.; Powers, J. C. *J. Biol. Chem.* **1984**, *259*, 4279.
22. Powers, J. C.; Boone, R.; Carroll, D. L.; Gupton, B. F.; Kam, C.-M.; Nishino, N.; Sakamoto, M.; Tuhy, P. M. *J. Biol. Chem.* **1984**, *259*, 4288.
23. Kaiser, E.; Colescott, R. L.; Bossinger, C. D.; Cook, P. I. *Anal. Biochem.* **1979**, *34*, 595.
24. Stein, P. D.; Hunt, J. T.; Floyd, D. M.; Moreland, S.; Dickenson, K. E. J.; Mitchell, C.; Liu, E. C.-K.; Webb, M. L.; Murugesan, N.; Dickey, J.; McCullen, D.; Zhang, R.; Lee, V. G.; Serafino, R.; Delaney, C.; Schaeffer, T. R.; Kozlowski, M. *J. Med. Chem.* **1994**, *37*, 329.
25. Yoshino, H.; Veda, N.; Nijima, J.; Sugumi, H.; Kotake, Y.; Koyanagi, N.; Yoshimatsu, K.; Asada, M.; Watanabe, T.; Nagasu, T.; Tsukahara, K.; Iijima, A.; Kitch, K. *J. Med. Chem.* **1992**, *35*, 2496.
26. Ellingboe, J. W.; Spinelli, W.; Winkley, M. W.; Nguyen, T. T.; Parsons, R. W.; Moubarak, I. F.; Kitzen, J. M.; Vonengen, D.; Bagli, J. F. *J. Med. Chem.* **1992**, *35*, 707.

Chapter 12

Libraries of Transition-Metal Catalysts

High-Throughput Screening of Catalysts for Synthetic Organic Methodology

Kevin Burgess, Destardi Moye-Sherman, and Alex M. Porte

Department of Chemistry, Texas A & M University,
College Station, TX 77843–3255

High throughput screening of catalysts is described. Batches of up to 96 different catalyst systems were established simultaneously, then screened sequentially. This technology has been applied to a C-H insertion and a cyclopropanation reaction; several interesting results emerged, notably silver-based optically active catalysts. The scope of high throughput screening is currently limited by the number of catalysts and catalyst precursors that are conveniently available. Therefore, synthesis of a new phosphine ligand type was performed, illustrating a scheme which can be adjusted to give many related ligands. It is anticipated that high throughput screens of such a series of ligands will facilitate fine-tuning of catalyst properties.

Most applications of combinatorial chemistry(*1-3*) so far have focused on identification of biologically active compounds. Very few uses have been reported outside this area. Libraries of materials have been screened for superconductivity,(*4,5*) and other reports have dealt with libraries of potential metal binders.(*6,7*) One paper describes how to make heterogeneous metal catalysts on a nanoscale.(*8*) However, despite considerable speculation, there is no report of effective combinatorial methods for finding catalysts for organic syntheses. Why should this be so? This paper considers that question, and suggests *high throughput screening of catalysts* may be a viable alternative.

High Throughput Screening of Catalysts.

At the advent of combinatorial chemistry, several new biotechnology companies set out to make huge libraries, typically via split syntheses or related techniques. More recently, the attention paid to large libraries has waned; indeed, some major players in the field have deserted them altogether in favor of one-compound-per-well approaches. The reason for this is that screening large libraries can be relatively difficult and/or ambiguous. An array with one compound per well therefore can be an attractive compromise between serial syntheses and screening at one extreme, and truly combinatorial chemistry at the other (Figure 1).

In catalysis, the state of the art is firmly entrenched in serial preparation and testing. There are several obvious advantages to this approach. In the context of

1054–7487/96/0128$15.00/0

homogeneous organometallic catalysts, for instance, the requisite complexes can be difficult to prepare so the very controlled environment of a one-at-a-time strategy has an obvious appeal. However, this mode of screening is relatively slow.

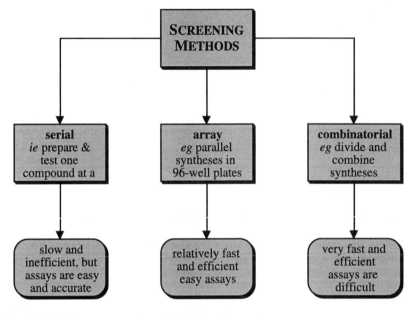

Figure 1. Different approaches to the task of screening.

Recently, there has been some discussion of *combinatorial catalysis*, although most papers using this term imply that it can be done, but do not report any examples. For instance, in one project, several mixtures of ligating groups and metals were applied to a functionalized polymer, and the resulting derivatized polymers were tested, one at a time, as catalysts in phosphate hydrolysis.*(9)* This type of approach is, in fact, standard in heterogeneous catalysis. Does the term "combinatorial catalysis" therefore also apply to decades of zeolite chemistry, for example? If it does, then there is nothing new about this type of combinatorial catalysis, and it is not useful for homogeneous systems. Another report uses the term "combinatorial catalysis" in the context of solid phase preparations of five related ligands which were then cleaved and tested sequentially.*(10)* In our view, the only genuine reports of combinatorial catalysis to date involve antibodies with genetically engineered variable regions to produce libraries of "abzymes".*(11,12)*

The word "combinatorial" means that substances must be combined at some stage in the process. The difficulties associated with preparation and screening of catalysts tend to be accentuated in any combinatorial process, so combinatorial catalysis is a more challenging proposition than preparation and screening of molecules for biological activities. Two features make it relatively hard to make libraries of catalysts and screen them for catalytic activities. First, to exhibit catalysis a compound must be sufficiently robust to participate in the reaction without becoming irreversibly changed, but not so stable as to be inert. This is not a ubiquitous characteristic. Second, the process of one molecule catalyzing the formation of another is inherently more difficult to observe than binding between two substances (*eg* a hapten and an antibody).

Libraries of catalysts conceivably could be prepared via split syntheses (one catalyst per bead); however, most catalytically active molecules are generally unsuited to solid phase syntheses, and there is very limited scope for solid phase syntheses of catalysts in which several components could be varied. Labile organometallic complexes certainly could not be efficiently prepared in this way. In some cases, catalysts could be combined and screened (as mixtures) to see which mixture mediates a given reaction, followed by a deconvolution process. This strategy is only suitable for situations in which the potential catalysts do not cross react, and establishing that this requirement is satisfied is a non-trivial undertaking. Alternatively, mixtures of compounds could be added to single potential catalysts to explore which of them are substrates, but analysis of the resulting mixtures becomes increasingly problematic as the number of substrates used is increased. We conclude that *combinatorial catalysis is possible, but there will be no approach that will have the generality and significance of the combinatorial methods currently used for syntheses and screening of biologically active molecules.*

We also have concluded that array methods are likely to compare very well with schemes for combinatorial catalysis. Arrays of one catalyst and one substrate per well are relatively easy to arrange. Many reactions could be run in parallel, so the efficiency of screening an array of catalyst systems is higher than in one-at-a-time approaches. Finally, analyses of each of the reactions can be done in just the same way as in serial screens, but the fact that many samples are generated simultaneously makes automation a time-saving option.

The following subsections describe our first attempts to realize high throughput catalyst screening using hardware which is currently available in many laboratories. The first transformation explored was a C-H insertion reaction.

High Throughput Screening of Catalysts for a C-H Insertion Reaction. The first reaction studied was Sulikowski's catalyzed intramolecular decomposition/C-H insertion reaction of diazocompound **1**.*(13)* Four chiral centers are formed simultaneously in this process, but two of them are irrelevant for the synthesis of mitosenes (the intended application of this reaction). A DDQ oxidation was used to remove the two irrelevant chiral centers giving the tricyclic system **3** for which the diastereomeric ratio is easily accessed by HPLC. Control experiments show that the menthyl ester alone has no significant effect on the stereoselectivity of the reaction; a chiral catalyst is required to form one diastereoisomer of this process preferentially (reaction 1).

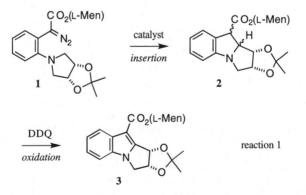

Figure 2 shows the approach used for the initial screening process. A microtitre plate was used to examine 96 potential systems for achieving the transformation of diazocompound **1** into **3**. Solvents and reagents were mixed in the plate, which was then gently agitated under an inert atmosphere for 18 h at ambient temperature. The plate was brought into the air, and the product was oxidized by adding DDQ into each well. Each reaction mixture was then filtered through a silica gel plug, an internal standard (naphthalene) was added, and their compositions were analyzed using an HPLC equipped with an autosampler. In this way, 96 sets of approximate yield and stereoselectivity data were obtained in less than a week. Finally, the most promising leads were reinvestigated in the standard way, *ie* sequentially.

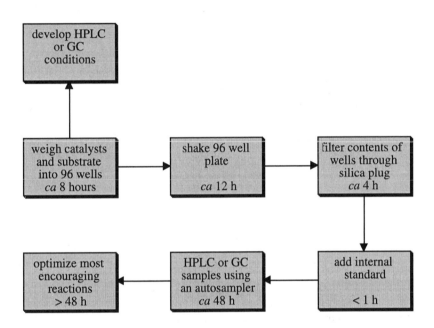

Figure 2. Procedure for high throughput screening.

Negative results (no reaction and/or poor stereoselectivities) in the screening process outlined above could be due to weighing errors or some unforeseen variable in the experimental protocol. Consequently, the screen cannot be used to reliably eliminate potential catalysts from further consideration. However, the screening process is ideal for identifying leads for further study.

The best catalytic systems identified after the screening/optimization process involved ligand **4** and copper (+1) triflate. This same ligand/metal combination was identified in Sulikowski's original work,(*13*) hence in one respect this observation marks a successful control experiment. However, THF was shown to be a better solvent than that originally used (CH$_2$Cl$_2$), so an improvement was also found. Consequently, the best diastereoselectivity observed was 3.9:1 as compared with 2.3:1 previously.(*13*) Perhaps more remarkable is the observation that ligand **4** and silver (+1) hexafluoroantimonate affords higher yields than the copper-based catalyst systems, and is almost as stereoselective. Use of silver complexes to mediate C-H insertion reactions is certainly uncommon, and perhaps unique. High throughput screening of catalysts definitely facilitated this discovery.

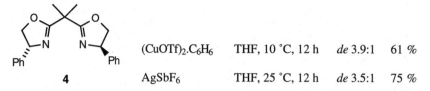

(CuOTf)$_2$.C$_6$H$_6$	THF, 10 °C, 12 h	*de* 3.9:1 61 %
AgSbF$_6$	THF, 25 °C, 12 h	*de* 3.5:1 75 %

4

The moderate stereoselectivities obtained in the screening process described above reflect the fact that there are four different reaction pathways by which the two diastereomers of **3** may form (Figure 3). Any catalyst that is intended to favor production of one of these isomers has to selectively reduce a specific combination of two of the four activation energies involved. Consequently, it may be difficult to obtain high diastereoselectivities in this transformation.

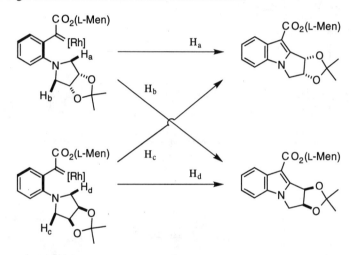

Figure 3. The origin of both stereoisomers in reaction 1.

High Throughput Screening of Catalysts for a Cyclopropanation Process.
Reaction 2 depicts a cyclopropanation process of interest in these laboratories.*(14)*

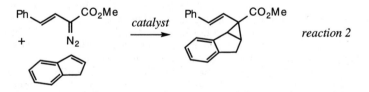

A plate of 96 different catalyst systems were screened as outlined above. The McKervey/Davies catalyst **5***(15,16)* gave the best results in the screen, and a moderately high enantioselectivity could be obtained after a brief optimization process. It is unsurprising that this particular catalyst was identified in view of the work by Davies on similar cyclopropanation reactions.*(16)* Interestingly, ligand **4** in conjunction with silver (+1), and with copper (+1), were the only other systems which gave appreciable enantioselectivities.

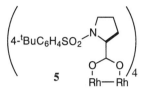

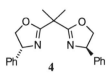

	(CuOTf)$_2$.C$_6$H$_6$	THF	0 °C	65 %	27 % *ee*
pentane, 0 °C; 68 %, 84 % *ee*	AgSbF$_6$	THF	25 °C	68 %	22 % *ee*

Conclusions Regarding High Throughput Screening of Catalysts. Array methods for screening catalysts occupy a useful middle-ground between serial screening and combinatorial catalysis. In any case, the latter is a mostly hypothetical proposition at present. The two sets of data given here represent libraries of catalysts focused on reactions of a reasonably well established type. Catalysts emerging from these screens include systems that might have been anticipated to perform well, and some unexpected hits too. In the next phase of our work we are screening transformations for which there is no known effective catalyst. Preliminary studies, to be reported soon, indicate that array methods can afford intriguing results in such cases.

Synthesis of a Series of Optically Pure C$_3$ Symmetric Ligands.

High throughput screens of transition metal complexes for catalytic activities require a relatively large number of complexes and/or ligands. The number of ligands that can be purchased is not vast, and the selection of optically pure ligands available is much smaller. For these reasons, syntheses in which a series of complexes or ligands can be easily obtained are valuable for high throughput screens, particularly if the products are optically pure. The final part of this manuscript describes one such synthetic route.

Phosphine ligands with C$_3$ symmetric, propeller shaped arrays of aromatic groups have structures which are both intriguing and unexplored with respect to asymmetric syntheses. Sharpless and co-workers made phosphines of this type, including **6**, but were unable to resolve them due to rapid equilibration of the enantiomers at room temperature.*(17,18)* None of the other C$_3$ symmetric phosphine ligands that have been reported have a propeller shape aromatic array (although some non-phosphine ligands do).*(19-30)*

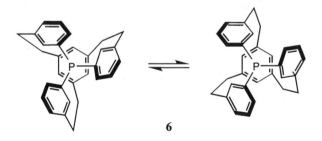

6

Our approach to generating a C$_3$ array of aromatic groups has been to attach a chiral substituent to the rings to induce one favored orientation over the other. Originally, our plan was to place these substituents *ortho* to the phosphorus atom. To this end, aryl bromide **7** was prepared via the following steps: (i) asymmetric reduction*(31)* of the corresponding ketone (one recrystallization gave optically pure material {GC}); and, (ii) methylation. Metalation of the aromatic ring with *tert*-butyllithium, followed by a PCl₃ quench, gave ligand **8**.

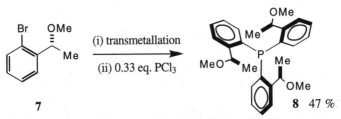

Unfortunately, ligand **8** is extremely crowded about the phosphorus atom, and is therefore not easily coordinated to a metal atom. Indeed, even oxidation of this phosphine to the corresponding phosphine oxide is extremely slow (50 °C, 14 h with H₂O₂ as compared with 10 min for PPh₃ under these conditions). Consequently, ligand **9**, the *meta* isomer of **8**, was prepared by an analogous method.

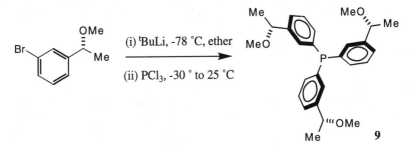

The phosphorus atom of ligand **9** is much more accessible than that of its isomer **8**. Figure 4 shows space filling diagrams of the two ligands, and this clearly illustrates the difference. Crystallographic data supports these models.

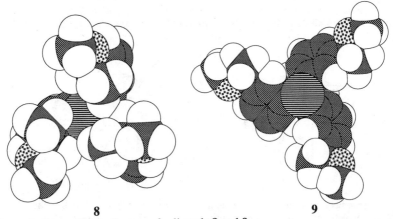

Figure 4. Space filling diagrams for ligands **8** and **9**.

Slight modifications to the synthesis of ligand **9** have facilitated preparation of several related ligands **10 - 14** for screening. Applications of high throughput screening techniques in conjunction with divergent ligand/catalyst syntheses will be reported soon.

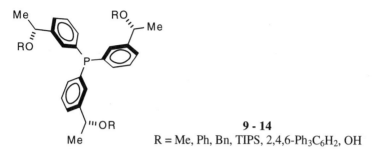

9 - 14
R = Me, Ph, Bn, TIPS, 2,4,6-Ph$_3$C$_6$H$_2$, OH

Literature Cited

1. Gallop, M. A.; Barrett, R. W.; Dower, W. J.; Fodor, S. P. A.; Gordon, E. M. *J. Med. Chem.* **1994**, *37*, 1233.
2. Gordon, E. M.; Barrett, R. W.; Dower, W. J.; Fodor, S. P. A.; Gallop, M. A. *J. Med. Chem.* **1994**, *37*, 1385.
3. Terrett, N. K.; Gardner, M.; Gordon, D. W.; Kobylecki, R. J.; Steele, J. *Tetrahedron,* **1995**, *51*, 8135.
4. Xiang, X.; Sun, X.; Briceno, G.; Lou, Y.; Wang, K.; Chang, H.; Wallacefreedman, W.; Chen, S.; Schultz, P. *Science* **1995**, *268*, 1738.
5. Briceno, G.; Chang, H.; Sun, X.; Schultz, P.; Xiang, X. *Science* **1995**, *270*, 273.
6. Burger, M. T.; Still, W. C. *J. Org. Chem.* **1995**, *60*, 7382.
7. Malin, R.; Steinbrecher, R.; Jannsen, J.; Semmler, W.; Noll, B.; Johannsen, B.; Frömmel, C.; Höhne, W.; Schneider-Mergener, J. *J. Am. Chem. Soc.* **1995**, *117*, 11821.
8. Müller, W. T.; Klein, D. L.; Lee, T.; Clarke, J.; McEuen, P. L.; Schultz, P. G. *Science* **1995**, *268*, 272.
9. Menger, F. M.; Eliseev, A. V.; Migulin, V. A. *J. Org. Chem.* **1995**, *60*, 6666.
10. Liu, G.; Ellman, J. A. *J. Org. Chem.* **1995**, *60*, 7712.
11. Chen, Y.-C. J.; Danon, T.; Sastry, L.; Mubaraki, M.; Janda, K. D.; Lerner, R. A. *J. Am. Chem. Soc.* **1993**, *115*, 357.
12. Janda, K. D.; Lo, C.-H. L.; Li, T.; III, C. F. B.; Wirsching, P.; Lerner, R. A. *PNAS* **1994**, *91*, 2532.
13. Lim, H.-J.; Sulikowski, G. A. *J. Org. Chem.* **1995**, *60*, 2326.
14. Burgess, K.; Ho, K.-K.; Moye-Sherman, D. *SYNLETT* **1994**, *8*, 575.
15. McKervey, M. A.; Ye, T. *J. Chem. Soc., Chem. Commun.* **1992**, 823.
16. Davies, H. M. L.; Hutcheson, D. K. *Tetrahedron Lett.* **1993**, *34*, 7243.
17. Bolm, C.; Sharpless, K. B. *Tetrahedron Lett.* **1988**, *29*, 5101.
18. Bolm, C.; Davis, W. D.; Halterman, R. L.; Sharpless, K. B. *Angew. Chem., Int. Ed. Engl.* **1988**, *27*, 835.
19. Burk, M. J.; Harlow, R. L. *Angew. Chem., Int. Ed. Engl.* **1990**, *29*, 1462.
20. Adolfsson, H.; Warnmark, K.; Moberg, C. *J. Chem. Soc., Chem. Commun.* **1992**, 1054.
21. Baker, M. J.; Pringle, P. G. *J. Chem. Soc., Chem. Commun.* **1993**, 314.
22. Bogdanovic, B.; Henc, B.; Meister, B.; Pauling, H.; Wilke, G. *Angew. Chem., Int. Ed. Engl.,* **1972**, *11*, 1023.
23. Kaufmann, D.; Boese, R. *Angew. Chem., Int. Ed. Engl.* **1990**, *29*, 545.

24.   Canary, J. W.; Allen, C. S.; Castagnetto, J. M.; Wang, Y. *J. Am. Chem. Soc.* **1995**, *117*, 8484.
25.   LeCloux, D. D.; Tokar, C. J.; Osawa, M.; Houser, R. P.; Keyes, M. C.; Tolman, W. B. *Organometallics* **1994**, *13*, 2855.
26.   LeCloux, D. D.; Keyes, M. C.; Osawa, M.; Reynolds, V.; Tolman, W. B. *Inorg. Chem.* **1994**, *33*, 6361.
27.   Nugent, W. A.; Harlow, R. L. *J. Am. Chem. Soc.* **1994**, *116*, 6142.
28.   Nugent, W. A. *J. Am. Chem. Soc.* **1992**, *114*, 2768.
29.   Ward, T. R.; Venanzi, L. M.; Albinati, A.; Lianza, F.; Gerfin, T.; Gramlich, V.; Tombo, G. M. R. *Helv. Chim. Acta* **1991**, *74*, 983.
30.   Burk, M. J.; Feaster, J. E.; Harlow, R. L. *Tetrahedron Asymmetry* **1991**, *2*, 569.
31.   Mathre, D. J.; Thompson, A. S.; Douglas, A. W.; Hoogsteen, K.; Carroll, J. D.; Corley, E. G.; Grabowski, E. J. J. *J. Org. Chem.* **1993**, *58*, 2880.

Chapter 13

Mixtures of Molecules Versus Mixtures of Pure Compounds on Polymeric Beads

Magda Stankova, Peter Strop, Charlie Chen, and Michal Lebl[1]

Selectide Corporation, 1580 East Hanley Boulevard, Tucson, AZ 85737

Two methods of screening, one-bead-one-compound synthetic strategy followed by on bead binding assay, and iterative screening in solution, were compared in assays designed for finding inhibitors of an enzyme of the coagulation cascade, factor Xa. It was found hat the results of iterative screening depend on the starting point of iteration, and may miss the best ligand contained in the library. The lead found by the iterative technique was structurally different from the inhibitors discovered by the one-bead-one-compound technique.

There are basically two methods for screening to exploit molecular diversity -- screening of individual compounds or screening of synthetic mixtures (for recent reviews on library techniques see e.g. (1-3), or the dynamic database on Internet (4)). Library technology based on the one-bead-one-compound method (5, 6) utilizes the first method for screening in the case of bead binding based screening. In this case each bead represents an individual compound and the screening result is independent of the remaining beads present in the mixture. Screening of one-bead-one-compound (OBOC) libraries for activities in solution requires partial release of the compound from the polymeric support, and is best performed with mixtures of gradually decreasing complexity (7, 8). Sreening in solution is most commonly an iterative approach (9, 10). In this case mixtures with defined features (one or two conserved building blocks, where the building blocks are amino acids for peptide libraries) are synthesized and building blocks leading to biological activity of the mixture are used as the starting point for the synthesis of the next generation library After several steps of iteration, the structure of the most active compound is defined. Advantages of mixture screening include: (i) independence of the analytical techniques, which are critical for successful one-bead-one-compound technique, (ii) no limitation in the number of compounds screened -- the one-bead-one-compound technique is clearly limited by the number of beads used for synthesis and screening, and, most importantly, (iii) independence of the biological test. In the list of disadvantages of screening of mixtures we can name: (i)

[1]Corresponding author

1054–7487/96/0137$15.00/0

their higher synthetic demands, (ii) the fact that independent leads will seldom be discovered, and (iii) the fact that a positive response may be the result of the synergy of several structures. One-bead-one-compound techniques are synthetically very simple, and all independent leads are revealed in a single screening step. However, not all biological tests can be modified for on-bead screening. In this manuscript we compare the results of two different approaches, for the screening on a target of real value -- factor Xa.

We have screened a number of one-bead-one-peptide libraries for ligands of factor Xa, and we have found a number of potent inhibitors of this important enzyme of the coagulation cascade (11). In addition we prepared an iterative library of tetrapeptides and acylated tripeptides (structure O$XX) on hydroxy-TentaGel, in which the first building block (O) was "defined" -- selected from 9 amino acids, 38 carboxylic acids and hydrogen. The second position ($) was "semi-defined" - - we have used 8 groups of five amino acids which were coupled as mixtures. The third and fourth position (X) was completely randomized -- by 21 and 23 amino acids respectively. We have therefore synthesized 384 pools of 2,415 compounds in each pool (927,360 compounds in total). The compound mixtures were cleaved from the carrier by 0.1 M NaOH, solution was neutralized by acetic acid, and the mixtures lyophilized. The pools were dissolved in 1 ml of screening buffer, generating a solution totaling 5 mM of screened compounds, in which each individual component was present at a concentration of 2.1 μM. At the same time we synthesized several mixtures containing peptide ligands defined earlier in a mixture format identical to the format of the iterative library as positive controls for the screening experiment.

Results of the screening of the library are given in Figure 1. The first column is the activity of the buffer with no added compound. Listed in columns 2 through 4 are the activities of synthetic mixtures containing three different inhibitors in the library format -- the positive controls for the screening experiment. The remaining columns present the results of seven groups (out of 48 groups defined by the last position) of mixtures. The groups shown here are those in which the mixture synthesized as positive controls should be included, together with those in which significant activities were found. For example, the third column (subgroup containing tyrosine, aspartic acid, lysine, arginine, and D-isoleucine) in group 11 (group defined by coupling acetyl as the last randomization building block), should contain the known lead Ac-Tyr-Ile-Arg in the mixture, and its activity should be comparable to the activity of the positive control mixture represented in column 4. As can be seen, activities comparable to the activities of control mixtures were found in groups 7, 10, and 11. Activities similar to those represented by groups 8 and 9 were not considered significant. Our original intention was to follow the iterative protocol of leads from groups 10 and 11 where the known ligands were expected, however, the activities of two subgroups from groups 18 and 46 prompted us to follow the alternative strategy, and deconvolute these, obviously more active, unexpected leads.

Two step deconvolution of the lead from group 46 gave the structure Aaa-Bbb-Ccc-Ddd (the full structure has not yet been released for publication) with Ki=0.65 μM. This activity is actually significantly lower than the activity of an unrelated lead that we have previously discovered, Ac-2Nal-Chg-Arg (Ki=0.27 μM), which was synthesized as one of components of the positive library mixtures (column 8 in group 11). This result clearly illustrates the danger inherent in the use of iterative techniques - the most active compound is not identified in the screening due to the fact that its most important

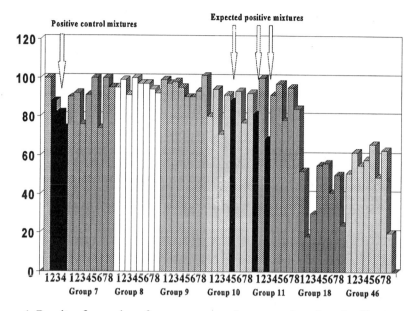

Figure 1. Results of screening of seven representative groups from iterative library. X axis -- controls and individual mixtures, for definition see text; Y axis -- % of response (hydrolysis of chromogenic substrate by factor Xa)

features are not defined in the early steps of iteration. The iterative pathway then follows the structural type which may not be optimal. We tested this hypothesis using model libraries having very different starting points for iteration. The results of the testing of model mixtures are given in Table I. Both the new structure, Aaa-Bbb-Ccc-Ddd, and the known ligand, Ac-2Nal-Chg-Arg, were used as the templates for synthesis of "left" and "right iteration" models. In these models the leftmost (or rightmost) building block of the structure was defined, the second bulding block from the left (or right) was semi-defined (mixture of five building blocks was used in this position), and the last two building blocks were completely randomized. As can be seen from the activities in the table, left-iteration model mixture of Aaa-Bbb-Ccc-Ddd is more active than left-iteration model mixture of Ac-2Nal-Chg-Arg. For the right-iteration model mixtures just the opposite is true. The left-iteration model mixture of Aaa-Bbb-Ccc-Ddd is three times more active than right-iteration model mixture of the same motif, whereas right-iteration model mixture of Ac-2Nal-Chg-Arg is three times more active than its left-iteration model mixture. Obviously, the critical residues in both motifs are on opposite sides of the molecule, and it is not surprising that iteration starting from the left side leads to the Aaa-Bbb-Ccc-Ddd motif, which would not be found with iteration starting from the right side (note that iteration from the right side is technically substantially more difficult than iteration from left side). The one-bead-one-compound method found the best motif due to the fact that individual structures rather than partially defined mixtures were used in screening. We can only speculate at this moment what would be the result of screening of the "library of

Table I. Activities of leads and model mixtures

Compound/Mixture	Structure	Activity[a]
OBOC lead	Ac-2Nal-Chg-Arg	$K_i = 0.27\ \mu M$
Left iteration model	Ac-(2Nal)-Xxx-Xxx	22 %
Right iteration model	Xxx-Xxx-(Chg)-Arg	45 %
Iteration lead	Aaa-Bbb-Ccc-Ddd	$K_i = 0.65\ \mu M$
Left iteration model	Aaa-(Bbb)-Xxx-Xxx	27 %
Right iteration model	Xxx-Xxx-(Ccc)-Ddd	11 %

[a] K_i value, or percent of inhibition at 100 nM concentration

libraries", which may represent the iterative library with all possible combinations of two or three building blocks (12). It would also be interesting to compare the results for positional scanning libraries (13), and for the newly described technique of non-iterative deconvolution, using multidimensional orthogonal mixtures of compounds prepared during robotic synthesis (Felder and Kris, IBC Symposium on Combinatorial Chemistry, Jan 24, 1996, San Diego, CA).

In conclusion, the iterative technique produced an alternative, structurally dissimilar inhibitor of factor Xa, which may prove valuable in the development of the new anticoagulant. The one-bead-one-compound method found the compound with the highest activity, yet may overlook alternative structures. Importantly, the results from iterative techniques depend on the position from which the iteration starts and may miss the most active compounds. It may be advisable to apply both screening techniques for those cases where the biological test is amenable to assay both by direct on-bead binding and by inhibition in solution.

Literature Cited
1. Ellman, J. A.; Thompson, L. A. *Chem.Rev.* **1996**, *96*, 555-600.
2. Rinnova, M.; Lebl, M. *Collect.Czech.Chem.Commun.* **1996**, *61*, 171-231.
3. Krchnak, V.; Lebl, M. *Molecular Diversity* **1996**, *1*, in press.
4. Lebl, M. Compilation of papers in molecular diversity field; INTERNET World Wide Web address: http://vesta.pd.com.
5. Lam, K. S.; Salmon, S. E.; Hersh, E. M.; Hruby, V. J.; Kazmierski, W. M.; Knapp, R. J. *Nature* **1991**, *354*, 82-84.
6. Lebl, M.; Krchnák, V.; Sepetov, N. F.; Seligmann, B.; Strop, P.; Felder, S.; Lam, K. S. *Biopolymers (Pept.Sci.)* **1995**, *37*, 177-198.
7. Salmon, S. E.; Lam, K. S.; Lebl, M.; Kandola, A.; Khattri, P. S.; Wade, S.; Pátek, M.; Kocis, P.; Krchnák, V.; Thorpe, D.; Felder, S. *Proc.Natl.Acad.Sci.USA* **1993**, *90*, 11708-11712.
8. Lebl, M.; Krchnák, V.; Salmon, S. E.; Lam, K. S. *Methods: A Companion to Methods in Enzymology* **1994**, *6*, 381-387.

9. Pinilla, C.; Appel, J.; Blondelle, S. E.; Dooley, C. T.; Dorner, B.; Eichler, J.; Ostresh, J.; Houghten, R. A. *Biopolymers (Pept.Sci.)* **1995**, *37*, 221-240.

10. Houghten, R. A.; Pinilla, C.; Blondelle, S. E.; Appel, J. R.; Dooley, C. T.; Cuervo, J. H. *Nature* **1991**, *354*, 84-86.

11. Seligmann, B.; Abdul-Latif, F.; Al-Obeidi, F.; Flegelova, Z.; Issakova, O.; Kocis, P.; Krchnak, V.; Lam, K. S.; Lebl, M.; Ostrem, J.; Safar, P.; Sepetov, N.; Stierandova, A.; Strop, P.; Wildgoose, P. *Eur.J.Med.Chem.* **1995**, *30 (supplement); Proceedings of the XIIIth International Symposium on Medicinal Chemistry; Muller,JC (Editor)*, 319s-335s.

12. Sepetov, N. F.; Krchnák, V.; Stanková, M.; Wade, S.; Lam, K. S.; Lebl, M. *Proc.Natl.Acad.Sci.USA* **1995**, *92*, 5426-5430.

13. Pinilla, C.; Appel, J. R.; Blondelle, S. E.; Dooley, C. T.; Eichler, J.; Ostresh, J. M.; Houghten, R. A. *Drug.Develop.Res.* **1994**, *33*, 133-145.

BIOLOGY-BASED CHEMICAL LIBRARIES

Chapter 14

Generation of Solution-Phase Libraries of Organic Molecules by Combinatorial Biocatalysis

Yuri L. Khmelnitsky[1], Peter C. Michels[1], Jonathan S. Dordick[2], and Douglas S. Clark[3]

[1]EnzyMed, Inc., 2501 Crosspark Road, Iowa City, IA 52242
[2]Department of Chemical and Biochemical Engineering, University of Iowa, Iowa City, IA 52242
[3]Department of Chemical Engineering, University of California, Berkeley, CA 94720

Combinatorial biocatalysis is a powerful methodology for synthesizing libraries of organic compounds in solution. Advantages of biocatalysis for generating organic libraries include the natural diversity of enzymatic reactions, the compatibility of reaction conditions and high-throughput screening techniques, ease of automation, and the ability to retrace synthetic pathways leading to active products. The integration of biocatalysis with high-speed robotics thus represents a new avenue of biotechnology for the discovery of new molecules and biotransformation schemes. The versatility of this approach is demonstrated by the synthesis of diverse libraries from small organic precursors, as well as the iterative derivatization of taxol, a complex natural product.

Combinatorial methods for drug discovery involve the assembly of various molecular building blocks in different combinations to produce libraries of new compounds. Syntheses of combinatorial libraries of both linear oligomers (such as peptides [1,2], oligonucleotides [3], or peptoids [4]) and non-oligomeric small organic molecules (e.g., benzodiazepines [5,6], hydantoins [6], or pyrrolidines [7]) have been reported (for recent reviews on combinatorial chemistry, see [8-12]). The majority of combinatorial synthetic work has been aimed at producing large libraries of compounds starting from biologically inert precursors. The libraries are then screened for biological activity, and prospective lead compounds are identified. In other words, the main objective of combinatorial synthesis is to generate new drug leads, and the recent progress in this area shows that combinatorial chemistry is well suited for lead discovery.

Finding a promising lead compound is still only a starting point on the long road to developing the actual drug. As a rule, the lead compound will not possess the

1054–7487/96/0144$15.00/0

complete combination of pharmacological properties required for a fully functional drug, such as high potency, low toxicity, and bioavailability. Additional derivatization of the lead compound is therefore needed in order to acquire these requisite characteristics. At this second stage of the drug discovery process, known as lead optimization or lead development, combinatorial approaches have been less effective. Moreover, conventional combinatorial strategies become more problematic as the lead molecule becomes more complex. Potential problems include difficulties associated with the reversible attachment of structurally diverse molecules to solid supports, the possibility of unfavorable side reactions, and detrimental effects of the harsh reaction conditions often employed in combinatorial synthetic schemes. These unfavorable factors become especially severe in the case of fragile natural product and complex synthetic leads.

It is thus evident that lead optimization can be a bottleneck in the drug discovery process, and there is a need to develop techniques that can rapidly and efficiently generate libraries of derivatives from lead compounds. To address this problem and extend combinatorial synthesis to lead optimization, we have devised the concept of combinatorial biocatalysis. Combinatorial biocatalysis uses chemical reactions catalyzed by enzymes and microorganisms to generate libraries of derivatives from existing lead compounds. Due to the inherent ability of enzymes to catalyze conversions of mixtures of complex molecules in solution--under mild conditions and without the formation of byproducts--combinatorial biocatalysis is ideally suited for creating libraries of derivatives from structurally diverse lead compounds.

In the present paper we describe several applications of combinatorial biocatalysis for the derivatization of small organic molecules, as well as a complex natural lead compound, taxol.

Combinatorial Biocatalysis

Combinatorial biocatalysis utilizes enzymatic reactions and microbial transformations to generate libraries of new derivatives from an existing lead. Reactions are performed iteratively; first-generation derivatives are modified by another round of biocatalytic reactions at additional reactive sites (either originally present in the starting compound or introduced in the first set of biocatalytic reactions) to produce a second generation of derivatives. By repeating this procedure it is possible to create numerous new derivatives of the original lead compound after several iterations.

An underlying principle of combinatorial biocatalysis is that in many cases enzymes are able to accept a wide variety of different molecules as substrates, provided these molecules have a functional group and/or structural element recognizable by the enzyme. A typical lead molecule contains several such functional groups, each amenable to reactions with several different enzymes.

A lead compound can be modified in at least two different ways. First, combinatorial biocatalysis takes advantage of the most common strategy for derivatization in combinatorial chemistry, namely the attachment of chemical building blocks to the lead molecule. Ester, amide, carbonate, carbamate, or glycoside linkages between the new moieties and the core structure can be efficiently formed enzymatically under mild conditions, with regio- and/or enantioselectivity. Second, the structural backbone of the starting compound can be altered by extending the carbon skeleton, or by introducing new (or modifying existing) functionalities on the molecule. As a result, new scaffolds for further derivatization are created. Examples of enzymatic reactions of this kind include carbon-carbon bond formation, hydroxylations, halohydrin formation, and oxidation or reduction of alcohols, aldehydes, ketones, and acids. These two approaches complement one another: new structures can be added to fresh reactive sites on the lead molecule created by enzymatic modification, and these new structural components can in turn be modified to provide additional sites for further derivatization.

The choice of biocatalytic conversions available for combinatorial biocatalysis is quite wide and continually growing due to ongoing advances in the field of applied biocatalysis. Numerous new reactions catalyzed both by isolated enzymes and microorganisms have been developed and applied in organic synthesis (for recent reviews on synthetic applications of biocatalysts, see [13-18]). The progress in cloning of novel enzymes is rapidly expanding the breadth of available biocatalytic transformations. Furthermore, recent advances in catalyst and medium engineering have resulted in the development of readily available improved biocatalysts, possessing high activity and stability under conditions previously regarded as unacceptable for enzymes, such as increased temperatures and non-aqueous environments [19-25].

Table I. Examples of Enzyme Reactions Available for Combinatorial Biocatalysis

Introduction of new functional groups	Modification of existing functionalities	Addition onto functional groups

Carbon-carbon bonds
- acyloin condensation
- cyanohydrin formation
- aldol condensation
- dimerization
Hydroxylation
Halogenation
Peroxidation
Halohydrin formation
Epoxidation
Diels-Alder cycloaddition
Addition of ammonia
Hydroperoxidation
Hydrogenation
	Oxidation of alcohols to
	  aldehydes/ketones
	Reduction of aldehydes/
	  ketones to alcohols
	Oxidation of sulfides to
	  sulfoxides
	Oxidation of aminogroups to
	  nitrogroups
	Oxidation of thiols to
	  thioaldehydes
	Hydrolysis of nitriles to
	  amides and carboxylic acids
	Replacement of aminogroups
	  with hydroxyl groups
	Lactonization
	Isomerization
	Epimerization
		Acylation
		  vinyl esters
		  trihaloethyl
		  esters
		  vinyl carbonates
		  vinyl carbamates
		  oxime esters
		  oxime carbonates
		  bifunctional
		
		Glycosylation
		  glycosides
		  aminoglycosides
		  glycosidic acids
		
		Amidation
		  amides
		  peptides

Examples of applicable biocatalytic reactions are summarized in Table I. These reactions are divided into three major categories, each involving chemical functionalities on the original or derivatized lead. Enzymes are thus viewed in terms of their reactivity toward particular functional groups, some subset of which is present on virtually all organic molecules. Thus, nearly all organic molecules are potential substrates for combinatorial biocatalysis.

Results and Discussion

Derivatization of Small Organic Molecules. The application of combinatorial biocatalysis for creating large diverse libraries from a single lead compound was demonstrated for three organic molecules of different structures, (±)-(2-endo,3-exo)-bicyclo[2.2.2]oct-5-ene-2,3-dimethanol (BOD), 2,3-(methylene-dioxy)benzaldehyde (MDB), and adenosine (ADS) (Figure 1). These compounds were selected because they have very different structures, which include several structural motifs common in compounds exhibiting biological activity, such as alicyclic and aromatic rings, heterocycles, double bonds, and hydroxyl groups. After identifying potential sites for enzymatic derivatization, the next step was to identify suitable biocatalysts (isolated enzymes or microorganisms) and reaction conditions. A limited optimization study to determine the most effective biocatalysts and favorable reaction conditions was then conducted. Based on these considerations, an iterative derivatization tree was constructed and used to guide the actual library synthesis.

Relevant biocatalytic reactions included oxidation, reduction, halohydrin formation, cyanohydrin formation, deamination, glycosylation, acyloin condensation, acylation, and microbial hydroxylation. For demonstration of these reactions, over 50 commercially available enzymes of different classes were used as catalysts. A typical reaction tree used for modification of MDB by combinatorial biocatalysis is shown in Figure 2. Similar schemes for BOD and ADS are presented in Figures 3 and 4, respectively. The reactions were carried out either in aqueous solutions or organic solvents, depending on the specific requirements of the reaction type (for example, enzymatic acylations must be done in nonaqueous media to suppress undesirable hydrolysis) or the solubilities of reagents. Each type of enzymatic modification was performed in a separate reaction vial and was catalyzed either by an individual biocatalyst or by a mixture of enzymes of similar type (for example, several lipases from different sources could be combined in the same vial to provide a mixed acylating catalyst). Formation of new derivatives was confirmed by HPLC or GC analysis of reaction mixtures.

It should be noted that each round of glycosylation or acylation reactions shown in Figures 2-4 can produce numerous derivatives depending on the number of glycosyl and acyl donors employed. For derivatization of BOD, MDB and ADS, three glycosyl and sixteen acyl donors were used. Some of the acyl donors were bifunctional, e.g. divinyl esters of dicarboxylic acids. When such donors were used, the starting compound itself was converted into an acyl donor (Acylation I in Figure 2), which was then used to acylate a diverse array of different alcohols (Acylation II in Figure 2). By attaching different moieties (sugars, acyl donors, and alcohols) in a combinatorial fashion, hundreds of highly diverse derivatives were quickly produced. Based on the number of new peaks appearing in chromatograms of reaction mixtures, the combined total of derivatives produced from BOD, MDB, and ADS was over 1800. Structures of some of the members of these libraries are shown in Figures 2, 5 and 6. The structures

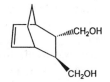

BOD

(±)-(2-endo,3-exo)-bicyclo
[2.2.2]oct-5-ene-2,3-dimethanol

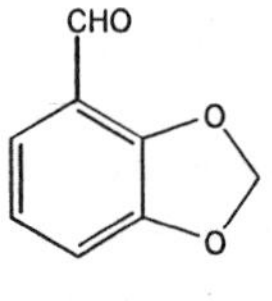

MDB
2,3-(methylenedioxy)
benzaldehyde

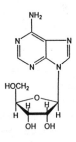

ADS
adenosine

Figure 1. Structures of BOD, MDB, and ADS.

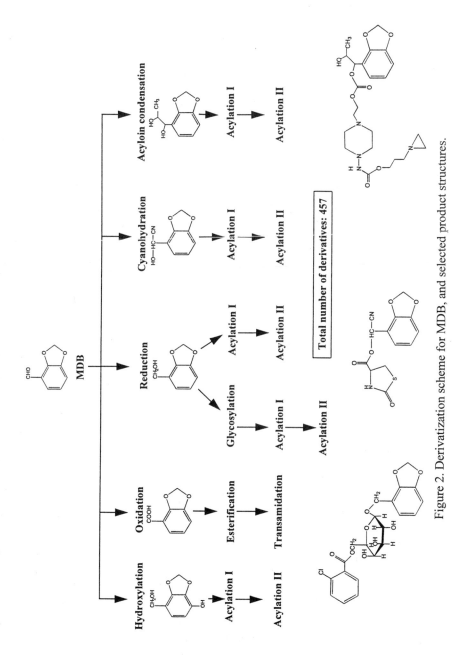

Figure 2. Derivatization scheme for MDB, and selected product structures.

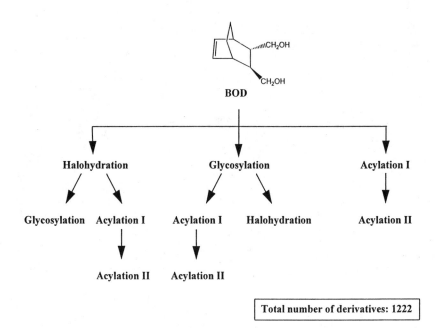

Figure 3. Derivatization scheme for BOD

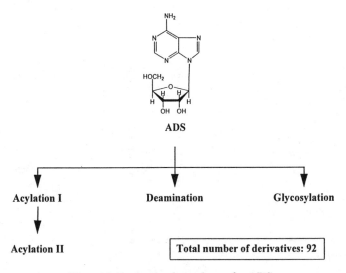

Figure 4. Derivatization scheme for ADS

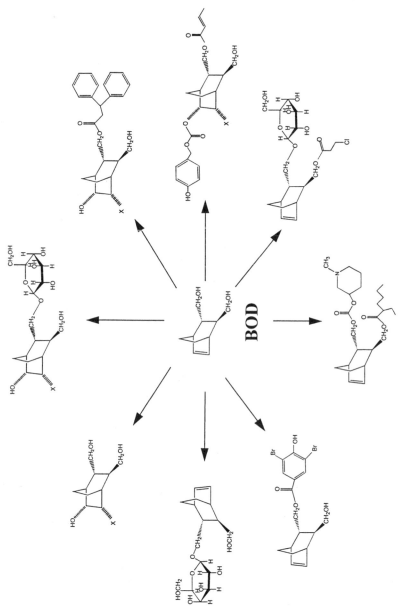

Figure 5. Structures of selected BOD derivatives.

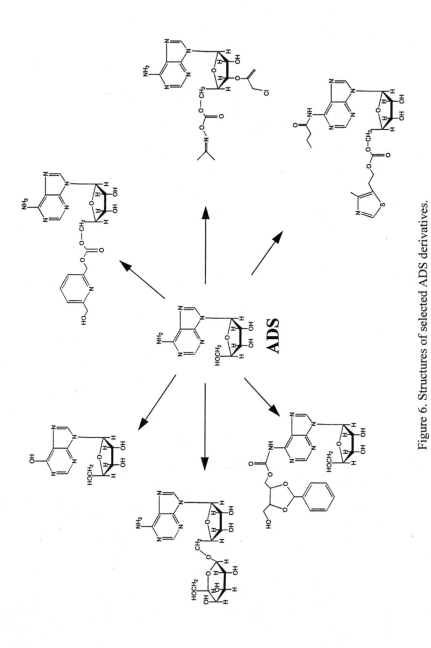

Figure 6. Structures of selected ADS derivatives.

of all derivatives produced during the first iteration of biocatalysis were confirmed by mass spectrometry and NMR.

Derivatization of Taxol

Combinatorial biocatalysis is by no means restricted to derivatization of relatively small organic lead molecules such as BOD, MDB and ADS. In fact, because of the diversity of biocatalysts available for solution-phase reactions, this strategy should be general to almost any lead compound, including complex natural products. This point is illustrated by the use of combinatorial biocatalysis to produce derivatives of taxol, a naturally occurring diterpenoid recently approved for the treatment of ovarian cancer [26]. As shown in Figure 7, taxol can be readily derivatized enzymatically, for example, by acylating hydroxyls in positions 2' and 7. By using combinatorial enzymatic acylations, similar to acylations I and II shown in Figure 2, a library of nearly 200 taxol derivatives was produced.

Several new derivatives of taxol from this library illustrate the potential for directing library synthesis towards derivatives with improved pharmacological properties. For example, a serious problem with taxol is its extremely low water solubility, which creates difficulties for administration of the drug. Several taxol derivatives, synthesized using combinatorial biocatalysis, showed greatly improved solubility in water. For example, a glucosylated taxol derivative is 60 times more soluble in water than taxol, and an acidic derivative is more than three orders of magnitude more soluble (Figure 7). In both cases, enzymatic modification of the 2'-hydroxyl group produced potential taxol prodrugs in the absence of undesirable byproducts.

Concluding Remarks

The results of this work clearly show that by using a wide variety of biocatalytic transformations, combinatorial biocatalysis can produce large libraries of derivatives from highly diverse lead compounds. Suitable leads include small organic molecules as well as complex natural products. Combinatorial biocatalysis is well suited for compounds dissolved in solution and in this respect differs from many other combinatorial technologies. This aspect of combinatorial biocatalysis allows for an effective interface between synthesis and screening. Thus, the biosynthetically produced libraries can be screened immediately without having to remove products from solid supports. Furthermore, the reactions in combinatorial biocatalysis represent single-step conversions that do not require extreme temperatures or corrosive reagents. Instead, they are performed under mild conditions that are very similar from one reaction to another. Uniform reaction conditions facilitate automation and iteration.

It should also be stressed that combinatorial biocatalysis is not limited to functional groups that are initially present on a given lead compound. For example, enzymes and microbial systems can be used to add functional groups to unactivated parts of an organic molecule. This capability is important because it enables significant changes in the basic structure of a lead compound, and expands the diversity of structures available from a small organic molecule. In addition, the natural

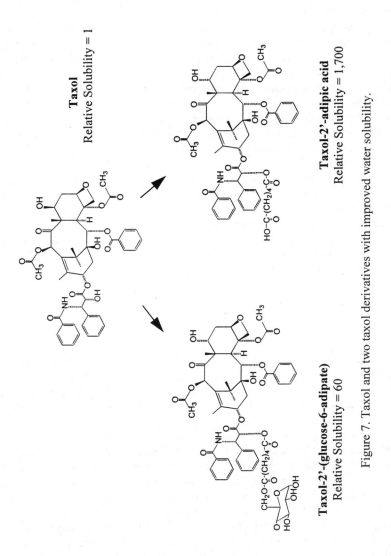

Figure 7. Taxol and two taxol derivatives with improved water solubility.

regioselectivity of enzymatic reactions provides the opportunity to create highly focused libraries, in which lead compounds are derivatized at specific sites in a controlled fashion.

Although the main objective of combinatorial biocatalysis is the discovery of new biologically active compounds rather than their large-scale production, in most cases the reaction sequence leading to the active compound can easily be scaled up to produce the active derivative on a preparative scale (grams to tens of grams). This built-in production capability is another important advantage of combinatorial biocatalysis.

Finally, the combinatorial synthetic strategy described here is by no means limited by purely biocatalytic reactions. In principle, it is readily adaptable to incorporate selected reactions from the vast arsenal of synthetic organic chemistry, at least in cases when these steps are compatible with the iterative biocatalytic scheme. In other words, combinatorial biocatalysis can be viewed as a chemoenzymatic approach to the synthesis of libraries, involving a fusion of enzymatic conversions, microbial transformations, and chemical reactions. In this context, it is worthwhile to mention that chemoenzymatic methods have, over the years, proven to be very effective in the synthesis and modification of numerous biologically active compounds [27].

General Experimental Approach

Biocatalytic reactions are carried out on a scale ranging from 0.5 ml to 100 ml, depending on the requirement of the isolated products for subsequent derivatization steps. A broad range of different solvents and their mixtures can be used as reaction media, depending on specific requirements of substrate solubility, enzyme stability and activity, and reaction thermodynamics. Due to the mild reaction conditions employed (temperatures ranging from 25°C to 55°C, atmospheric pressure, and mild stirring), standard glass vials and flasks suffice as reaction vessels.

Normally, each reaction vessel houses a single type of enzymatic conversion, although several enzymes or reagents of the same kind can be used to carry out the reaction. For example, a mixture of several lipases or proteases can be used as a catalyst instead of an individual enzyme, or several similar building blocks can be added into the same vial to be attached to a reactive site on the lead molecule. Upon completion of a reaction step in the iterative derivatization scheme, further enzymatic conversions can often be carried out in the same reaction vial without any work-up of the reaction mixture or isolation of the product. The biocatalyst can be removed from the reaction mixture by simple filtration (for whole cells and insoluble or immobilized enzymes) or by high speed ultrafiltration (for soluble enzymes).

Reaction mixtures are analyzed by HPLC or GC, and the number of synthesized derivatives is estimated from the number of new peaks appearing on chromatograms of reaction vials compared with non-enzymatic controls. Many of the derivatives obtained have been further characterized by GC/MS and ^{1}H-NMR. Derivatives of the lead compound synthesized as single compounds or as mixtures may be submitted to an appropriate screening protocol for determination of biological activity. Libraries showing a positive response in a biological screen can be further subjected to a recursive deconvolution procedure in order to identify the active compound that manifests the desired biological activity. Every library of derivatives is characterized by the series of biocatalytic reactions used to produce it, a so-called "biocatalytic history". The biocatalytic history of each library is unique and can be retraced, yielding the specific reaction sequence leading to the active product. The structure of the active compound can then be deduced, based on knowledge of the enzymatic reactions employed (the high specificity and predictability of enzymatic reactions makes this step possible), and confirmed by standard analytical techniques.

Literature Cited

1. Salmon, S. E.; Lam, K. S.; Felder, S.; Yeoman, H.; Schlessinger, J.; Ullrich, A.; Krchnak, V.; Lebl, M. *Acta Oncologica* **1994**, *33*, 127.
2. Furka, A. Drug Dev. Res. 1995, 36, 1.
3. Sherman, M. I.; Bertelsen, A. H.; Cook, A. F. Bioorg. *Med. Chem. Lett.* **1993**, *3*, 469.
4. Simon, R. A.; Kania, R. S.; Zuckermann, R. N.; Huebner, V. A.; Jewell, D. A.; Banville, S.; Ng, S.; Wang, L.; Rosenberg, S.; Marlowe, C. K.; Spellmeyer, D. C.; Tan, R.; Frankel, A. D.; Santi, D. V.; Cohen, F. E.; Bartlett, P. A. *Proc. Natl. Acad. Sci. USA* **1992**, *89*, 9367.
5. Bunin, B. A.; Plunkett, M. J.; Ellman, J. A. *Proc. Natl. Acad. Sci. USA* **1994**, *91*, 4708.
6. DeWitt, S. H.; Schroeder, M. C.; Stankovic, C. J.; Strode, J. E.; Czarnik, A. W. *Drug Dev. Res.* **1994**, *33*, 116.
7. Murphy, M. M.; Schullek, J. R.; Gordon, E. M.; Gallop, M. A. *J. Am. Chem. Soc.* **1995**, *117*, 7029.
8. Madden, D.; Krchnak, V.; Lebl, M. Perspect. *Drug Disc. Des.* **1995**, *2*, 269.
9. Terrett, N. K.; Gardner, M.; Gordon, D. W.; Kobylecki, R. J.; Steele, J. *Tetrahedron* **1995**, *51*, 8135.
10. Gordon, E. M.; Gallop, M. H.; Campbell, D.; Holmes, C. P.; Bermak, J.; Look, G.; Murphy, M.; Needels, M.; Jacobs, J.; Sugarman, J.; Chinn, J.; Fritsch, B. R. *Eur. J. Med. Chem.* **1995**, *30*, S337.
11. Gallop, M. A.; Barrett, R. W.; Dower, W. J.; Fodor, S. P. A.; Gordon, E. M. *J. Med. Chem.* **1994**, *37*, 1233.
12. Gordon, E. M.; Barrett, R. W.; Dower, W. J.; Fodor; S. P. A.; Gallop, M. A. *J. Med. Chem.* **1994**, *37*, 1385.
13. Jones, J. B. *Tetrahedron* **1986**, *42*, 3351.
14. Crout, D. H. G.; Christen, M. *Mod.Synth.Meth.* **1989**, *5*, 1.
15. *Applied Biocatalysis*; Blanch, H.W.; Clark, D.S., Eds.; Marcel Dekker: New York, 1991.
16. Wong, C. H.; Shen, G. J.; Pederson, R. L.; Wang, Y. F.; Hennen, W. J. *Meth. Enzymol.* **1991**, *202*, 591.
17. Turner, N. *J. Nat. Prod. Rep.* **1994**, *11*, 1.
18. Theil, F. *Chem. Rev.* **1995**. *95*, 2203.
19. *Biocatalysis at Extreme Temperatures. Enzyme Systems Near and Above 100 °C*; Adams, M.W.W.; Kelly, R.M., Eds.; ACS Symp. Ser. 498; American Chemical Society: Washington, DC, 1992.

20. Martinek, K.; Mozhaev, V. V. In *Thermostability of Enzymes*; Gupta, M. N., Ed.; Springer-Verlag: Berlin, 1993; pp 76-83.
21. Khmelnitsky, Yu. L.; Levashov, A. V.; Klyachko, N. L.; Martinek, K. *Enzyme Microb. Technol.* **1988**, *10*, 710.
22. Dordick, J. S. *Enzyme Microb. Technol.* **1989**, *11*, 194.
23. Klibanov, A. M. *Acc. Chem. Res.* **1990**, *23*, 114.
24. Randolph, T. W.; Blanch, H. W.; Clark, D. S. In *Biocatalysts for Industry*; Dordick J. S., Ed.; Plenum Press: New York, 1991; pp 219-237.
25. Dordick, J. S. *Biotech. Progr.* **1992**, *8*, 259.
26. Holmes, F. A.; Kudelka, A. P.; Kavanagh, J. J.; Huber, M. H.; Ajani, J. A.; Valero, V. ACS Symp.Ser. 583; American Chemical Society: Washington, DC, 1995; 31.
27. Parida, S.; Dordick, J.S. *J. Am. Chem. Soc.* 1991, 113, 2253.

Chapter 15

Living Libraries Using Gene Transfer: A Renaissance for Natural Products

K. A. Thompson

Department of Molecular Biology, ChromaXome Corporation,
11111 Flintkote Avenue, San Diego, CA 92121

Despite millennia of human effort, 99% of microbes and plants on Earth cannot be grown in a laboratory or tilled field. The 1% that have been grown generated half of the top-selling drugs in 1994. The same 1% also produced actinomycins, esperamicins, camptothecin, and a wide range of anti-infective and anti-cancer agents. What unseen treasures could be produced by the missing 99%? This chapter will describe the rationale for developing an alternative method for accessing natural products, especially the missing 99%, as well as the technological basis for the potential of this approach, and the resulting new opportunities--such as "un-natural" natural products.

Rationale for Pursuit of Natural Products

What is a Natural Product? Technically, all organic chemicals made by living organisms are natural products, whether primary or secondary metabolites. However, the parlance of the pharmaceutical industry reflects a narrower definition, centered around secondary metabolites. The focus of a typical natural product chemist is on the specific subset of natural products, also called simply "small molecules", with a molecular weight of up to 1000 daltons. This subset of chemicals is frequently responsible for the pharmacological activity of natural product samples (*1*). Small molecules can be generated by synthetic chemists as well, but the richest diversity and complexity has been seen in small molecules from natural products. This chapter will utilize the narrower definition of natural products.

What is Not a Natural Product? One critical point is often not appreciated: it takes more than one gene to make a natural product. Most of the biologicals intended for use as therapeutics are proteins or peptides. These are generated by expression of single gene products, or by direct peptide synthesis. In contrast, biosynthesis of natural products (in the sense indicated above) requires a suite of enzymes, often pre-assembled as a metabolic pathway, as shown in Table I.

1054–7487/96/0158$15.00/0

Table I. Origins of Selected Biochemicals

Natural Products		Biologicals (2)	
Name	*Number of Genes (approx.)*	*Name*	*Number of Genes*
Puromycin (*3*)	4	DNase	1
Tetracyclines (*4*)	12	Interleukin 2	1
Doxorubicin (*5*)	9	Factor VIII	1
Bialaphos (*6*)	5	Erythropoietin	1

Consequently, discovery strategies utilizing molecular biology and gene transfer entail different considerations for natural products (i.e., multi-gene processes) than for biologicals (i.e., single gene products).

It is important to understand this distinction when interpreting the range of manipulations possible using molecular biology tools, as discussed below.

Value of Natural Products. The desirable properties of natural products reflect their active roles in nature (*7*). They are chemical weapons for sessile organisms, providing both offense and defense for filamentous soil organisms (prokaryotic and eukaryotic), and soft, vulnerable, marine sponges, for example. They are external signaling molecules for organisms in aqueous milieux, providing communication between or within communities of unicellular organisms (*8, 9*). These biological activities have established natural products as essential contributors to the pharmacopoeia, as shown in Table II (*10;* Leach, R. E., University of California, San Diego, personal communication, 1995). Half of the group listed below is derived directly or indirectly from natural sources, through extraction, a combination of extraction and synthetic modification, or purely synthetic production based on the natural product structure.

The properties of natural product-based drugs are due in part to the unique resource available in live cells: enzymes. Enzymes can perform stereospecific synthesis, generating chiral centers. They function in the presence of extraordinarily complex mixtures, with a daunting efficiency. They routinely perform biosynthetic steps and choreographies that have not been imagined by the most clever and innovative synthetic chemists. Depending on the extracellular milieu, enzymes perform synthetic steps with an extraordinary range of reagents. Marine organisms, for example, take advantage of the presence of halogens in seawater, incorporating them into a variety of unique metabolites.

Table II. Origin of 25 Top-Selling Drugs in 1994

	Natural Product		*Synthetic Chemical*
Trade Name	*Chemical Class*	*Original Source Organism*	*Trade Name*
Premarin Oral	Steroid	Mammal	Zantac
Amoxil	Beta-lactam	Fungus	Procardia
Synthroid	Amino Acid	Mammal	Vasotec
Trimox	Beta-lactam	Fungus	Prozac
Lanoxin	Glycoside	Plant	Cardizem CD
Augmentin (Clavulanate)	Beta-lactam	Bacterium	Proventil Aerosol
Augmentin (Amoxicillin)	Beta-lactam	Fungus	Coumadin Sodium
Mevacor	Terpene	Fungus	Ventolin Aerosol
Amoxicillin Trihydrate	Beta-lactam	Fungus	Cipro
Zestril	Amino Acid	Bacterium	Zoloft
Provera	Steroid	Plant	Capoten
Acetaminophen/Codeine	Alkaloid (Codeine)	Plant	Propoxyphene N/APAP
Ceclor	Beta-lactam	Fungus	Dilantin

Problems of Natural Product Commercialization. Despite this appeal, the majority of the large pharmaceutical houses no longer test natural product samples from the wild. Why? To start, there is a significant chance that they will not be able to harvest or culture enough of a promising lead to completely test the drug in clinical trials, much less to go to market. It may seem romantic to harvest a potent seaweed from the high seas, or tree bark from primeval forests. However, depending on potency and the yield during extraction, tons of raw material may be required just to take a drug lead through pre-clinical testing. For some species, or for a sub-population producing a key chemical, a harvest on that scale would consume the entire global biomass. Furthermore, the biomass required for ongoing production of a drug in the market can be tons per week.

Grimmer yet, and only recently realized, is that only a small fraction of fungi, bacteria, and plants can be grown under the controlled conditions that would be needed for commercialization, despite decades of concerted academic and industrial effort (*11, 12*). As illustrated in Figure 1, the current natural product repertoire is derived predominantly from the cultivable minority of organisms.

Figure 1. Natural Products from Cultivable and Non-Cultivable Sources

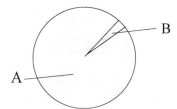

A. From Uncultivable Organisms	*B. From Cultivable Organisms*	
	Actinomycins	Vincristine
?	Penicillins	Cephalosporins
	Aminoglycosides	Quinine
	Erythromycin	Tetracyclines
	Amphotericin	

Plants generally have larger biomass than microbes; as a result it has been possible to sample plant natural product chemistry in the absence of sustained plant culture. In contrast, it has not been possible to sample the chemistry of uncultivable micro-organisms, with a few notable exceptions.

For example, marine sponges act as natural incubators of microbes, since half of their biomass may be comprised of only two or three species of symbiotic bacteria. The chemistry of some symbiotic microbes has been inadvertently examined, and until recently believed to belong to the sponge (*13*). The products are unique and have unexpected structures.

For microbes, then, little is known about the chemistry of uncultivable organisms other than isolated cases. However, with the recent advent of the polymerase chain reaction (PCR), it has become clear that vast numbers of species-- estimates range from 90 to 99.9% of total biodiversity--are present, but have never been cultured, counted, or analyzed (*14*). PCR is a rapid and sensitive, yet robust, method for amplifying molar amounts of DNA. These features have made PCR an excellent tool for the detection of new species based on the presence of trace amounts of unique chromosomal DNA sequences in complex environmental samples (*15, 16*).

Synthesis, the alternative to growth of the natural product source, is often not feasible. Unfortunately, features such as chirality make natural products effective at the molecular level, but also make them difficult to synthesize *in vitro*. A well-known case is taxol, a potent anti-tumor agent originally found in yew trees and more recently in a fungus (*17*). In nature, the synthesis is performed by enzymes, and each chiral center is generated by stereospecific enzymes. Taxol has been synthesized *in vitro*, despite its multiplicity of chiral centers. However, the final yield is too inefficient to be commercially competitive.

Another difficulty of natural products is the complexity of extracts. Even if a hit is detected, it can be difficult to isolate the responsible component. The complexity also complicates or obscures interpretation of assays, since multiple activities may act in concert.

Finally, there are the additional difficulties of culturing marine natural products--those of the 1% that can be cultured. Saline fermentation, for example, leads to corrosion of metal incubation chambers and fittings, and the inconstancy and inaccessibility of ocean waters require formulation of complex artificial seawater.

Need for a Renaissance in Natural Products. What is the future of natural products? The hurdles described above are unyielding, and are compounded by the diminishing returns of the remaining natural products programs in the big pharmaceutical houses. There is one glaring option: we need to access the missing 99%.

But how? One option is to convince the 99% to grow. This has been the focus of intense effort over the previous five decades, but there are no spectacular breakthroughs on the horizon.

An alternative, or a supplement, is to use gene transfer as illustrated in Figure 2 to bypass the need to culture. Tools and methods are available now to transfer genetic material into well-characterized, cultivable organisms, generating living libraries (*18, 19*).

Figure 2. Gene Transfer

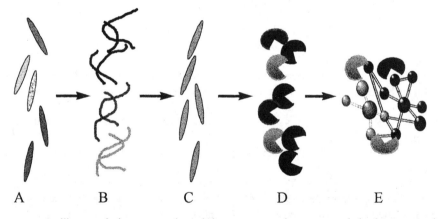

A   B   C   D   E

As illustrated, donor organisms (A) are processed to remove their chromosomal DNA (or in some cases mRNA) (B). The organisms do not need to be homogenous or actively growing. The DNA is transferred into recipient organisms (C), such as *E. coli* or *Streptomyces lividans*, and thereafter each recipient of a unique DNA fragment is described as a clone. In some cases, the DNA inserted into a clone will cause the production of one or more encoded proteins (D). In prokaryotes, genes for metabolic pathways are frequently located in clusters, increasing the likelihood of finding interacting proteins encoded on the same DNA fragment.

When those proteins are enzymes, and are functional, they can act independently or in concert with the pre-existing enzymes in the recipient organism.

The result in a small subset of clones is a change or supplement to the metabolism of the recipient (E).

A variety of available assays can be used to screen the resulting engineered organisms. The development of this alternative approach requires some investment of time and resources. However, the costs of making and screening living libraries must be weighed against the escalating economics and diminishing returns of conventional natural product discovery.

The potential is staggering: if the metabolic pathways of uncultivable organisms could be recovered with even the poor efficiency of 1%, this would double the global amount of natural products that would be available for screening. Furthermore, the integration of gene transfer tools into natural products discovery opens up a large set of benefits not previously available.

Added Value Using Gene Transfer Tools. A significant benefit of gene transfer is elimination of the need to culture the native organism, as discussed. This separates the enzymatic and regulatory demands of the specific and desired metabolic activities from the nutritional and reproductive needs of the intact organism.

Once isolated in living libraries, genes and pathways can be further manipulated for optimization, for example by gene shuffling (*20*). This manipulation can increase the specificity or tolerance of individual enzymes within a pathway. Individual enzymes or partial pathways can also be mobilized to modify extant pathways.

Furthermore, isolated genes and pathways are patentable. This provides an additional and not insignificant layer of proprietary protection for a biosynthetic process. This advantage is not as broadly available to synthetic chemists.

The size of the transferred genetic material in an individual sample can be between 0.1-10% of the starting genome. This reduces the complexity of extracts simply by reducing the number of metabolic pathways (beyond that provided by the host) that are present.

The individual samples in a living library are encoded due to the presence of unique DNA sequences, and thus can be pooled during the assay process. The DNA in each sample contains the production system, the information needed for identifying the cognate live organism, or for identifying the pool of organisms from which it came.

This method for capturing biosynthetic activity should not be seen as a competitor for synthetic chemistry, but as a complementary method. Each approach has significant advantages that can be exploited when suitable.

Limitations of Gene Transfer. One potential problem with the transfer of metabolic pathway genes is that currently available expression systems may not be representative of the genera contained in the missing 99%. Established expression systems are, after all, cultivable species.

For some metabolites, the cognate pathway may be larger than can be handled by current expression vectors. Oxytetracycline biosynthesis requires greater than 100 kilobases of genomic DNA (*21*). However, some expression systems are able to handle DNA fragments of this size, including BAC vectors in *E. coli* (300 kb capacity) and YAC vectors in yeast (1000 kilobases) (*22, 23*).

Genes transferred into a heterologous host may suffer from problems of foreign expression, such as unsuitable codon usage, or inappropriate regulatory sequences (for example, transcription termination sequences, or translation start sequences).

However, since the living libraries are renewable resources, each library can be re-examined or re-manipulated as new tools become available or are improved.

Need for Ultra-High-Throughput Screening.   Clearly, even if the described limitations can be ameliorated by advances in molecular biology understanding and tools, it will be necessary to examine large pools of engineered cells to detect desirable samples. This is illustrated by calculating the number of clones needed to represent a single prokaryotic genome (*24*). This number is based on the size of the source genome, the size of the genome fragments, and the desired thoroughness of the representation. For example, exhaustive analysis of a single average *Streptomyces* genome (8 megabases), converted into 35 kb fragment-containing clones, will require the examination of up to 2700 clones.

Thus an environmental library made from 1 gram of soil, which may contain 4000 distinct species (*25*), would generate hundreds of thousands of clones. It is essential to remove inactive and irrelevant samples from libraries of engineered cells to reach minimal acceptable values for individual samples. The minimal value required for a sample is determined by the cost of conventional high-throughput screening. A sample must have a threshold probability of generating a hit in a conventional panel of assays, or it is not economical to screen. This probability can be maximized by pre-screening with a panel of simpler and less expensive assays to dereplicate and to detect broad or relevant activities.

Technology

Combinatorial Biology is a Toolbox for Living Libraries.   Since gene cloning began to blossom in the 1970's, the availability and sophistication of expression systems has increased tremendously. New developments in expression systems in eukaryotes and prokaryotes have increased not only the ease of using these systems, but also their effectiveness.   This expansion is due in part to the need by the biotechnology industry to have production systems for single cloned enzymes and therapeutic proteins.

These events spawned the field of combinatorial biology, where molecular biologists use their tools in strategies familiar to combinatorial chemists. At the simplest level, individual genes undergo combinatorial modification. A gene acts in place of the chemist's chemical scaffold, and undergoes the limited modifications that are acceptable: switching of nucleotides between A, G, C, and T. The gene structure can also be rearranged, or "shuffled", so that the protein domains are reorganized or reshaped.

More complex modifications can be performed on entire pathways. First, they can act in place of the chemist's scaffold, and be subject to modification by random alteration, or addition. This could entail addition of single or multiple genes to a known pathway, acting as a simple derivatization of the pathway and its natural product (*26*). Second, the manipulations of a pathway can be a purely combinatorial mixing of genes, with no parallel to the chemist's scaffold. This provides one opportunity not available to combinatorial chemists--the chance to find truly new chemistry, not pre-defined or constrained by scaffolds.

At the core of combinatorial biology, the wealth of expression systems has followed the trend in basic research from the sequencing of genes toward the study of the cognate protein function. Ten years ago, the sequencing of a gene was a publishable event, but no longer. Now it is necessary to determine some function of the cognate protein, often by expression of the gene in hand, or a panel of its mutant descendants. This is facilitated by the variety of companies producing and supplying innovative and high-quality reagents and tools.

The expression systems in demand reflect the range and demands of the gene studies: eukaryotic (mouse, *Drosophila*, zebrafish and insect cells or whole organisms, fungi, and yeast) or prokaryotic (*E. coli*, *B. subtilis*, or *Streptomyces*) systems, inducible or constitutive expression, and products that are secreted, targeted to the nucleus, soluble or insoluble. Vectors carry the DNA into the expression host for extra-chromosomal propagation or insertion into the host genome, and are a key part of an expression system. Modifications to the vector provide the finest levels of tailoring and control over the system, even controlling the size of the DNA fragments that can be successfully transferred into the host.

The mechanics of assembly of these DNA sources and the suitable expression systems have similarly become more available and increased in quality, including enzymes, reagents, and methods and equipment for manipulation of megabase (>1000 kilobase pair) DNA. Sophisticated electrophoretic apparatuses for reliable analysis of megabase DNA have only recently become commercially available. Although relatively expensive, they are user-friendly and programmable.

The enzymes used to "cut and paste" DNA with exquisite specificity are becoming cheaper, cleaner, and greater in number and specificity. In particular, ongoing product development by the biotechnology supply companies has provided a flow of new "rare-cutter" enzymes. These enzymes cut DNA at specific sites that can be exceedingly rare, depending on the DNA composition of the targeted species. Thus these enzymes are ideal for manipulating megabase DNA from different branches of the tree of life.

Reagents and methods for DNA manipulation have also mushroomed as enabling tools such as PCR open new vistas of research. Scientists from a broad range of unrelated fields--such as oceanography, paleoentomology, archaeology, and forensic pathology--have been encouraged by the ease of performing and analyzing PCR reactions, and consequently have integrated DNA analysis into their work. Thus we now know how to recover DNA from ocean core samples, insects in amber, prehistoric granaries, and mummies. While mummy or dinosaur DNA may not be of immediate interest for a living combinatorial library, other than in movies, all of these sources also contain the remnants of microbes that may indeed be of interest.

Unique Tools Essential for Living Libraries of Metabolic Pathways. The tools described above are routinely used for the analysis and manipulation of single genes, or for manipulation of large fragments of genomic DNA for genome studies.

However, analysis and manipulation of multi-enzyme pathways requires additional tools.

Our laboratory initiated work on a combinatorial expression system to address the key problem with discovery of eukaryotic metabolic pathways: with few exceptions, their genes do not reside in physical clusters. Furthermore, without exception, eukaryotic genomes contain introns. Consequently, tracing the steps in a metabolic pathway is an excruciating process of stepping from one enzyme to the next, by complementation, mutant analysis, two-hybrid analysis of cDNAs, or other linear methods.   Our efforts are focused instead on a combinatorial approach to reconstitution or reassembly of pathways. Since our end goal is to find novel and diverse chemistry, it is to our benefit that the majority of combinatorial pathways will be "un-natural". (The issue of screening the resulting astronomical numbers of these clones will be addressed below.)

The generic design of the expression constructs is shown in Figure 3. This represents the combinatorial assemblage of DNA fragments that is introduced into each recipient cell.

Figure 3. Combinatorial Gene Expression Strategy

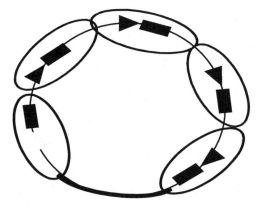

This system is comprised of a combinatorial concatemer of multiple DNA cassettes (ovals, 5 in this example), each acting as an independent cistron. Each cassette contains a relevant promoter (triangles), a control or experimental DNA fragment (rectangles), and a transcriptional terminator (not shown). The experimental DNA fragment can be prokaryotic genomic DNA or eukaryotic cDNA.

This combinatorial system will be less efficient than the capture of extant clustered pathways from the genomic DNA of prokaryotes. However, there are few alternatives for retrieving pathways from most eukaryotes.

Tools Available for Ultra-High-Throughput Screening. It is not enough to simply generate these libraries; they also need to be screened. This is the focus of a significant portion of our laboratory.

A variety of extant screens and analytical formats are compatible with the biological conditions necessary for living libraries. The conditions described as "biocompatibility" vary, depending on the expression system used and the specifics of the library being expressed. In general, however, biocompatibility requires a suitable pH (near neutral) and temperature (25-37 degrees C), as well as appropriate moisture, time, nutritional requirements and oxygenation.

For compatibility with living libraries, there are three main requirements. First, the key event of the screen must be biocompatible. This should not be an insurmountable hurdle since most questions addressed by screening are biological in nature. Second, the expression host used in the library must not trigger a positive response in the screen. Third, the method for reporting the key event must be biocompatible.

Established screens may at first glance appear to be unsuitable for integration with biological libraries, but upon closer examination can be re-engineered. Analysis of an assay response, is frequently performed by incorporation and detection of radioisotopes. The same assay may be converted to a second format that instead produces fluorescence, thus allowing its use in ultra-high-throughput screening of living libraries.

Compatible assays must also be converted to higher-efficiency formats. Our laboratory utilizes encapsulation, which has been used in industrial fermentation for growth and maintenance of production cultures (*27*). Individual cells are encapsulated in droplets of an inert, permeable matrix, followed by cultivation in a liquid medium.

The droplets can be viewed as miniature Petri dishes. The permeable matrices that are available are resilient, withstanding vortexing, extended fermentation, centrifugation, and growth of the encapsulated micro-colonies. Assay reagents can be added to the liquid medium, or crosslinked to the matrix prior to encapsulation. Culture media can be changed to favor the living library or a specific assay or target organism. Small droplets (20-100 microns) can be sorted by fluorescence-activated flow sorting. Larger droplets (>400 microns) can be sorted visually when chromogenic, luminescent, or fluorescent assay reagents are used. After sorting, the droplets can be dissolved under specific conditions to release the contents for further study or growth.

Flow cytometry is used as a diagnostic and research tool, but is under-utilized as a tool for examining large sample pools. High speed flow sorters can examine and sort particles (in this case cells or small encapsulated cells) at rates exceeding 26,000 particles per second (*28*). For a library of 50,000,000 clones, this translates into 40 minutes of sorting time. Since fluorescence is the signal detected, the procedure is not destructive, and the negative samples can be recycled to examine other conditions or assays. Multiple assays can in fact be performed in a single pass, by designing the assays to be simultaneous, producing either identical or distinguishable fluorescent signals. Some flow sorters will also tabulate and track the assay result(s) for each positive, allowing fingerprinting of each isolate. Furthermore, flow sorters can deliver positive cells directly into 96-well or 384-well plates, at defined numbers of particles per well.

These components have been assembled and are in use in our laboratory to integrate ultra-high throughput and bio-compatible sorting of living natural product libraries.

Opportunities

Access to the Missing 99%. What classes of chemicals have not yet even been imagined? What will be in that 99%? The field of natural products chemistry, reliant on the need to culture organisms, is in a position similar to that of protein chemists in the early 1900's. At that time, the protein chemists made substantial strides with proteins that could be extracted in large quantities and adequately purified. Proteins such as actin, hemoglobin, and the collagens were there for the taking, starting with a trip to the local abattoir.

But these proteins were only the tip of the protein iceberg. They were proteins that could withstand crude extraction methods, home-made reagents, and were superbly abundant. Collagens, for example, which comprise approximately one third of the total protein in a mammal, were extracted from bones with phosphoric acid (*29*).

Who in 1930 imagined the underlying clever strategies of transcription factors, the delicate but powerful webs of signal transduction pathways? This elegance remained invisible. This entire underworld of gene regulation and interactive signaling only began to be revealed when new tools became available: cloning, gene expression, monoclonal antibodies, *in vitro* cell culture, immunofluorescence, two-hybrid systems (*30*), and the network of support companies providing high-quality, time-saving commercial reagents.

Rare but critical proteins may comprise 0.01% or less of cellular protein, and only be present in restricted temporal and developmental locations, gender and species. These are now routinely "found" because of sequence homologies or delicate and transient interactions with other proteins. The protein itself is often never actually purified from tissue, but is detected only indirectly, using DNA probes or antibodies. Functional analysis and large scale production of the protein are routinely done in heterologous systems because of the intractability or inaccessibility of the natural source.

There are multiple parallels between those early protein chemists and the current state of the art in natural products drug discovery. Samples are derived from large quantities of biomass. Crude extracts are examined for recoverable activity or characteristic features. Fractionation is necessary to isolate the individual component of interest. The process is labor-intensive and tedious, and involves a significant level of artistry. Samples that are abundant, stable under the standard extraction methods, and consistently produced will dominate. Large scale purification requires substantial supplies of biomass and can retain impurities.

Thinking of these parallels between early protein chemistry and natural product drug discovery, the latter would be expected to overlook biochemicals with the following features:

◊ less abundant
◊ less stable with current extraction methods
◊ inconsistently or transiently produced
◊ inconsistently or transiently bioactive (i.e., present but requiring activation)
◊ present only in biomass not available in large quantities
◊ difficult to purify

Natural product drug discovery is now in the position to take advantage of the tools developed by and for the protein chemists, and to take them a step further. These tools may open another level of sensitivity, access, and control for natural product chemists.

Re-Mining the 1% Already in the Laboratory. Even for the relatively few cultivable organisms, metabolite production can be dramatically influenced by culture conditions. Some microbes, such as actinomycetes and fungi, undergo stages of differentiation that generate entirely different chemical and metabolic profiles. As a result, even cultivable organisms have not yet revealed their entire spectrum of metabolites. The integration of biosynthesis and bioassay as described in this chapter can thus also be applied to cultures and stocks currently in the banks of natural product discovery groups.

Mixing Pathways. For millennia, farmers have struggled to make hybrids between different species, hoping to benefit from a melding of distinctive features: flavor, color, scent. The biology of the living subjects imposed some restraints, however. It was not possible to blend the color of a flower with the flavor of a good mushroom, other than by cooking. However, if the color is produced by a polyketide or carotenoid pathway, and the flavor is generated by a combination of cyclic peptide biosynthetic pathways, "un-natural" natural products could be generated if you were able to mix the pathways in a live organism. This type of trans-species blending will provide an unlimited focus for experimentation and discovery as more pathways become available for gene manipulation.

Derivatization. In combinatorial chemistry, a chemical backbone is exposed to a matrix of modifications. Similarly, in combinatorial biology, a chemical backbone (within a living host) can be modified by a matrix of enzymatic modifications (within the same host). Cloned partial or complete metabolic pathways for metabolites such as polyketides, phenazines, aflatoxins, tetrapyrroles, or cobalamin (*31-35*) are available to act as backbones, while individual enzymes are available for performing derivatization (*36-39*).

Conclusion

The complexity and richness of natural products are unsurpassed. The challenge is to make them substantially more accessible. Access to natural products has historically been limited to the small fraction of our global biodiversity for which cultivation conditions have been determined. A strategy with staggering potential is to use gene transfer to generate living libraries to capture metabolic pathways from nature. Moreover, when integrated with ultra-high-throughput screening, this strategy also has the potential to become commercially competitive. As these strategies develop in our laboratory and others', it will be exciting to see what chemistry--natural and un-natural--is revealed in that missing 99%.

Acknowledgments

I would like to acknowledge the following scientists for helpful discussions and input regarding ongoing work: Michael Dickman, Janice Thompson, Nicole Nasby, and the ChromaXome Scientific Advisory Board, in particular Carl Djerassi, Michael Pirrung, Margo Haygood, and Mark Erion.

Literature Cited

1. Newman, D. J. *Soc. Indust. Microbiol. News* **1994**, *vol 44*, pp 277-283.
2. *The Merck Index*; Budavari, S.; O'Neil, M. J.; Smith, A.; Heckelman, P. E., Eds.; Eleventh Edition; Merck & Co., Inc.: Rahway, NJ, 1989.
3. Lacalle, R. A.; Tercero, J. A.; Jimenez, A. *EMBO J.* **1992**, *vol 11*, pp 785-792.
4. McDowall, K. J.; Doyle, D.; Butler, M. J.; Binnie, C.; Warren, M.; Hunter, I. S. In *Genetics and Product Formation in Streptomyces*; Baumberg, S., Ed.; Plenum Press: New York, NY 1991; pp 105-116.
5. Grimm, A.; Madduri, K.; Ali, A.; Hutchinson, C. R. *Gene* **1994**, *vol 151*, pp 1-10.
6. Raibaud, A.; Zalacain, M.; Holt., T. G.; Tizard, R.; Thompson, C. J. *J. Bacteriol.* **1991**, *vol 173*, pp 4454-4463.
7. Colwell, R. R. *Science* **1983**, *vol 222*, pp19-24.
8. Beppu, T. *Gene* **1992**, *vol 115*, pp 159-165.
9. Kaiser, D.; Losick, R. *Cell* **1993**, *vol 73*, pp 873-885.
10. *The Merck Index*; Budavari, S.; O'Neil, M. J.; Smith, A.; Heckelman, P. E., Eds.; Eleventh Edition; Merck & Co., Inc.: Rahway, NJ, 1989
11. Bull, A. T.; Goodfellow, M.; Slater, J. H. *Annu. Rev. Microbiol.* **1992**, *vol 46*, pp 219-252.
12. Mann, J.; Davidson, R. S.; Hobbs, J. B.; Banthorpe, D. V.; Harborne, J. B. In *Natural Products, Their Chemistry and Biological Significance*; Longman Scientific and Technical: Essex, England, 1994.
13. Faulkner, D. J.; He, H.; Unson, M. D.; Bewley, C. A. *Gazetta Chimica Italiana* **1993**, *vol 123*, pp 301-307.
14. Torsvik, V.; Goksoyr, J.; Daae, F. L. *Appl. Env. Microbiol.* **1990**, *vol 56*, pp 782-787.
15. Zhou, J.; Bruns, M. A.; Tiedje, J. M. *Appl. Env. Microbiol.* **1996**, *vol 62*, pp 316-322.
16. Liesack, W.; Stackebrandt, E. *J. Bacteriol.* **1992**, *vol 174*, pp 5072-5078.
17. Stierle, A.; Stroebel, G.; Stierle, D. *Science* **1993**, *vol 260*, pp 214-216.
18. Sambrook, J.; Fritsch, E. F.; Maniatis, T. *Molecular Cloning, A Laboratory Manual*; Cold Spring Harbor Laboratory Press: Plainview, NY, 1989.
19. Hopwood, D. A., et al *Genetic Manipulation of Streptomyces, A Laboratory Manual*; John Innes Foundation: Norwich, England, 1985.
20. Stemmer, W. P. *Proc. Nat. Acad. Sci.* **1994**, *vol 91*, pp 10747-10751.
21. McDowall, K. J.; Doyle, D.; Butler, M. J.; Binnie, C.; Warren, M.; Hunter, I. S. In *Genetics and Product Formation in Streptomyces*; Baumberg, S., Ed.; Plenum Press: New York, NY 1991; pp 105-116.
22. Shizuya, H.; Birren, B.; Kim, U.-J.; Mancino, V.; Slepak, T.; Tachiiri, Y.; Simon, M. *Proc. Nat. Acad. Sci.* **1992**, *vol 89*, pp 8794-8797.
23. Burke, D. T.; Olson, M. V. *Meth. Enz.* **1991**, *vol 194*, pp 251-270.
24. Clarke, L.; Carbon, J. *Cell* **1976**, *vol 9*, pp 91-99.
26. Hopwood, D. A.; Malpartida, F.; Kieser, H. M.; Ikeda, H.; Duncan, J.; Fujii, I.; Rudd, B. A.; Floss, H. G.; Omura, S. *Nature* **1985**, *vol 314*, pp 642-644.
27. Chibata, I.; Tosa, T. *Trends Biol. Sci.* **1980**, *April*, pp 88-90.

28. van den Engh, G.; Stokdijk, W. *Cytometry* **1989**, *vol 10*, pp 282-293.
29. *The Merck Index*; Budavari, S.; O'Neil, M. J.; Smith, A.; Heckelman, P. E., Eds.; Eleventh Edition; Merck & Co., Inc.: Rahway, NJ, 1989
30. Fields, S.; Song, O. *Nature* **1989**, *vol 340*, pp 245-246.
31. Hopwood, D. A.; Sherman, D. H. *Annual Review of Genetics,* Annual Reviews, Inc., 1990; Vol. 24, pp 37-66.
32. Pierson, L. S.; Thomashow, L. S. *Mol. Plant-Microbe Interactions* **1992**, *vol 5*, pp 330-339.
33. Yu, J.; et al *Appl. Env. Microbiol.* **1995**, *vol 61*, pp 2365-2371.
34. Hansson, M.; Hederstedt, L. *J. Bacteriol.* **1992**, *vol 174*, pp 8081-8093.
35. Raux, E.; et al *J. Bacteriol.* **1996**, *vol 178*, pp 753-767.
36. Beck, J.; Ripka, S.; Siegner, A.; Schiltz, E.; Schweizer, E. *Eur. J. Biochem.* **1990**, *vol 192*, pp 487-498.
37. Aubert-Pivert, E.; Davies, J. *Gene* **1994**, *vol 147*, pp 1-11.
38. Lees, N. D.; Skaggs, B.; Kirsch, D. R.; Bard, M. *Lipids* **1995**, *vol 30*, pp 221-226.
39. Hugueney, P.; et al *Plant J.* **1995**, *vol 8*, pp 417-424.

Chapter 16

Phage Expression of a De Novo Designed Coiled Coil Stem Loop Miniprotein Scaffold for Constrained Peptide Library Display

Robert M. Miceli[1,2,5], David G. Myszka[2,6], Catherine E. Peishoff[3], and Irwin M. Chaiken[1,2,4]

[1]Rheumatology Division, University of Pennsylvania School of Medicine, 913 Stellar Chance Labs, 422 Curie Boulevard, Philadelphia, PA 19104 Departments of [2]Molecular Immunology and [3]Physical and Structural Chemistry, SmithKline Beecham Research and Development, 709 Swedeland Road, King of Prussia, PA 19406

A synthetic cDNA for the coiled coil stem loop (CCSL) miniprotein was assembled and expressed as a fusion to gene III minor coat protein on the surface of filamentous phage. Phage containing the CCSL sequence were affinity-purified via an immunoaffinity tag built into the construction. An isolated phage clone, containing the expected CCSL sequence and designated CCπ.1, demonstrated binding to integrin receptor GPIIb/IIIa through the Arg-Gly-Asp sequence within the loop of the CCSL. Integrin receptor binding of CCπ.1 was selectively competed with a soluble Arg-Gly-Asp-containing peptide. CCSL expression on phage opens the way for construction of constrained helix and loop libraries which could prove useful in identifying novel peptide ligands for receptors, defining key side chains of native proteins involved in binding and establishing structural models for their pharmacophores. A Factor Xa site engineered into the construct provides a means to isolate and characterize CCSL variants selected from libraries following cleavage and immuno-purification of the CCSL from the phage particle. A molecular model of CCπ.1 miniprotein scaffold, deduced by homology modeling to a published crystal structure of a designed protein containing an antiparallel coiled coil, is also presented. This model will 1) help predict spatial orientations of coiled coil residues in CCSL variants 2) assist α-helical and loop library design and 3) facilitate 'grafting' of binding motifs from native proteins.

[4]Corresponding author
[5]Current address: Gull Laboratories, 1011 East 4800 South, Salt Lake City, UT 84132
[6]Current address: Department of Oncological Sciences, University of Utah School of Medicine, Salt Lake City, UT 84132

1054–7487/96/0172$15.00/0

Structural elements important for biological recognition are often associated with motifs built from α-helices. For example, growth hormone, IL2, and IL4 bind to their receptors via residues arrayed predominantly on the surface of α-helices (*1-3*). Additionally, proteins such as repressors and transcription factors mediate their specificity in binding to DNA through α-helical recognition sites (for review, see *4*). Similar relationships between loop-containing domains and binding recognition exist. For instance, the affinity and specificity of antibody variable regions is mediated through variations in loop length and amino acid sequence (*5*). Hence, mimetics of α-helical and loop surfaces represent a potential means for antagonizing native helix- and loop-presenting proteins.

One approach to designing mimics and antagonists which block biological processes is to use stable *de novo* designed structural units. These units can act as scaffolds for presentation of key protein residues or motifs similar to those found in the native state. Unfortunately, single-stranded α-helices are not typically stable in aqueous solution and require the additional stabilization of lower temperature or less polar solvents (*6, 7*). Recent efforts have instead centered on the development of alternative structural units for α-helical mimicry, including four-helix bundles and coiled coils. In contrast to single-stranded chains, helical arrays such as the coiled coil are highly stable in aqueous solution (*8, 9*).

Coiled coils exist naturally in many proteins, notably those whose functions are associated with structural integrity, such as keratin, myosin, and fibrinogen (*10*). Further, DNA-binding elements, members of the "bZIP" class of eukaryotic transcriptional activator, contain coiled coil motifs referred to as "leucine zippers" (*11, 12*). Coiled coils are formed by multiple α-helices aligned in parallel (or antiparallel) and in register that cross at angles of approximately 20° (*10*). These right-handed α-helices, when wound around one another in a left-handed supercoil, maximize the interchain packing of core hydrophobic and ionic residues responsible for maintaining orientation and stability (*10, 13, 14*). The amphipathic nature of the coiled coil is dictated by the biochemical properties of heptad repeat residues, denoted *abcdefg* (*15*) for one helix and *a'b'c'd'e'f'g'* for the other. For example, residues at *a* and *d* are commonly hydrophobic, lie on the same side of the helix, and make interior side-to-side contacts with residues at *a'* and *d'* of the opposing helix (*14*). In contrast, residues at positions *b*, *c*, and *f* are generally exposed to the solvent and vary more freely to provide surface specificity. Flanking residues at *e* and *g* are often of opposite charge and can form interhelical salt bridges (*10, 15*). In addition, their side chain methylene groups make intrahelical hydrophobic contacts. The side chain of *e* packs against the preceding *d'* of the adjacent chain and *g* against the following *a'* (*14*). These interactions together make important contributions to the stability of the coiled coil.

Successful *de novo* designs involving two-stranded a-helical coiled coils (*8, 9, 16-18*) and four-helix bundles (*19-22*) have been reported. Data taken from these experiments together with other studies comparing the amino acid sequences of heptad repeats from numerous native leucine zippers (*12, 23*) and coiled coil fibrous proteins (*24*) suggest that the coiled coil motif is tolerant to amino acid substitutions as long as changes conform to the general hydrophilic-hydrophobic profile of the heptad.

Recently, we designed and synthesized the first antiparallel coiled coil with an intervening loop sequence which we refer to as a coiled coil stem loop (CCSL,

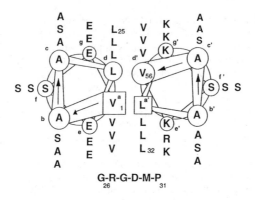

Figure 1. Helical wheel projection of the CCSL miniprotein. Leucine residues at position d in the amino-terminal α-helix interact across the interface with residues at position a' in the carboxy-terminal helix.

Figure 1) (*25*). This prototype CCSL was shown to be monomeric, helical and stable (*26*). We are currently interested in utilizing this CCSL prototype as a presentation scaffold for constrained libraries of helical and loop residues. Toward this end, we have undertaken effort to generate CCSL-constrained helical and loop libraries on filamentous bacteriophage. Previous work has shown that libraries of peptides (*27-29*), hormone (*30*) and enzyme variants (*31, 32*), and antibody variable regions (for review see *33*) can be successfully displayed using phage. Phage display of the CCSL miniprotein scaffold should provide versatility in manipulating structural motifs, including incorporation of peptide libraries at strategic points within the unit, and ease in mutagenesis of key residues. In addition, phage expression renders the ability to quickly screen large numbers of candidate ligands and enrich for binding species via relatively simple iterative screening techniques.

Here we describe the successful design, cDNA construction, and display on phage of a *de novo* designed CCSL miniprotein. We also report a working molecular model of our CCSL generated by comparison to an antiparallel coiled coil in a solved peptide crystal structure. This model will aid us in the design of constrained libraries and CCSL variants

Materials and Methods

Oligonucleotides.

Deoxyoligonucleotides were synthesized in an Applied Biosystem Model 394 DNA synthesizer (Foster City, CA) utilizing phosphoramidite chemistry. Oligonucleotides were purified using oligonucleotide purification cartridges (OPC). Deoxynucleotide cyanoethyl phosphoramidites, DNA synthesis columns (40 nM polystyrene) and OPCs were purchased from Applied

Biosystems. Crude oligonucleotide preparations were Mermaid purified (Bio101) prior to use in PCR. Oligonucleotide sequences:

CCSLIACV.1: 5' GATGGCCGACGCGGCCGTTGCCGCTCTAG AAAG-CGAGGTTAGTGCCCTTGAGAGCGAAGTGGCAAGCCTAGAAAGCGAG GTCGCGGC 3';

CCSLI.2: 5' TTGCCTTTACGGCACTCAACTTACTCTTAA CGGCAGC-TAATGGCATATCGCCTCGACCGAGTGCCG CGACCTCGC 3' ;

CCSLIACV.3SW:5'AATCAGCGACCGCTGCCAGCTTGCTCTTGACCG AAGCCA ATTTTGCCTTTACGGCAC 3';

CCSLI.4: 5' GATGGCCACAGCGGCCGCTGCTCT TCCCTCGATTGC-GGCATCGTCATCGTCCTTGTAATCAGC GCACGCTG 3'.

Construction rationale: Primer design and of the synthesis of CCSL miniprotein cDNA.

Oligonuleotides were designed for PCR synthesis of a synthetic cDNA encoding a 56 amino acid *de novo* designed coiled coil stem loop miniprotein that we have previously described and characterized (*26*). Using an approach described by Di Donato et al. (*34*), a "core template", containing the first 130 nucleotides of the synthetic gene, was constructed by gene splicing by overlap extension (SOE) (*35*), using oligonucleotides CCSLIACV.1 and CCSLI.2 (Figure 2). Successive PCR extensions, using CCSLIACV.3SW and

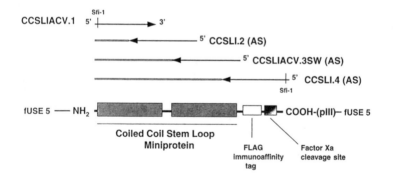

Figure 2. Assembly scheme for CCSL phage prototype cDNA. Synthesis of the cDNA for the CCSL prototype was performed by sequential PCR of four oligonucleotides encoding the coiled coil stem loop, FLAG immunoaffinity tag and the Factor Xa cleavage site. See *Materials and Methods* for details. Anti-sense primers are designated (AS). Boundary Sfi-I cloning sites are indicated and utilized for insertion of the cDNA into Sfi-I cut fUSE5 phage vector. Boxed regions represent the approximate alignments of the expressed prototype components to the oligonucleotides.

CCSLI.4, and exploiting engineered overlaps (T_m 46-48°C) at the 3' terminus, permitted extension of the "core template" to the final 247 nucleotide size. In addition to the synthetic CCSL, the construct was engineered to encode the FLAG immunoaffinity epitope (DYKDDDD) and a Factor Xa cleavage (IEGR) site as depicted in Figure 1. The final cDNA template was electrophoresed on a 1% agarose gel, purified by GeneClean (Bio101), and digested with restriction endonuclease Sfi-I (Boehringer Mannheim). Following gel and GeneClean purification again, the Sfi-I-cut CCSL/FLAG/Factor Xa insert was ligated to Sfi-I-cut fUSE5 vector overnight at 15°C. The fUSE5 vector and associated *E.coli* host strains were generously provided by Dr. George Smith, University of Missouri, Columbia, MO (*27, 36*). Competent DH10B cells (Gibco/BRL) were transformed using the ligation mix (2 µl of 20µl), and transformants were grown overnight in LB containing tetracycline (20µg/ml). Phage were PEG precipitated from the supernatant of these transformed cells and used for selection on anti-FLAG-coated oxirane beads (described below).

Affinity selection of CCSL phage.

Enrichment of FLAG-expressing CCSL phage was performed on anti-FLAG peptide monoclonal antibody (M2) (Kodak/IBI)-coated oxirane acrylic beads (Sigma) as previously described (*37*). Bound phage were eluted with 0.5 ml of 0.1M glycine, pH 2.5 for 10 minutes at room temperature with gentle rocking. Following elution, the acid eluate was neutralized with 50 µl Tris-HCl, pH 9.5 and concentrated to 200-300µl with a Centricon-30 microconcentrator (Amicon). Concentrated phage from the eluate were used to infect freshly-prepared "starved" *E.coli* K91-Kan cells and amplified overnight in LB (Luria both) containing 20 µg/ml tetracycline. Amplified phage were concentrated via sequential PEG (polyethylene glycol) precipitation and suspended in TBS (tris buffered saline, 50 mM Tris pH 7.5, 150mM NaCl) for further use. Sequencing-grade single-stranded DNA templates were prepared from individual clones isolated from these affinity-enriched populations (*38*).

DNA sequencing.

Sequencing was performed on a 373A automated DNA sequencer (Applied Biosystems) using the DyeTerminator or DyePrimer Cycle Sequencing chemistries.

Antibodies and Phage ELISAs.

Microtiter plates (Immunlon 3, Dynatech) were coated with either 100µl of anti-FLAG M2 (Kodak/IBI) or sheep anti-M13 phage antibody (5'Prime-3'Prime) (both 5 µg/ml) in 0.1 M sodium bicarbonate, pH 8.5 at 4°C overnight. Plates were washed five times with TBS/0.5% Tween 20 (v/v) before blocking with Blotto (5% nonfat dry milk (Carnation) in TBS, containing 0.02% NaN3) for 45 minutes at 4°C. Plates were washed and incubated with PEG-purified phage ($1x10^8$/well) diluted in TTB (TBS, 0.5% Tween (v/v), 1 mg/ml BSA (bovine serum albumin)) for 45 minutes. Following the binding phase, plates were washed 5 X with TBS/Tween and incubated with an appropriate dilution of horseradish peroxidase-coupled anti-phage conjugate (Pharmacia) in TTB for

45 minutes. All antibody and conjugate incubations were performed at 4°C. Plates were again washed extensively and developed with ABTS substrate (2,2'- Azino*bis* (3-ethylbenzothiazoline-6-sulfonic acid) diammonium salt, Kirkegaard and Perry). Plates were read on a ELISA plate scanner (Molecular Devices) at 405 nm after 10-15 minutes.

CCSL phage: RGD competition binding assay to GPIIb/IIIa.

Microtiter plates (EIA II Plus, Linbro) were coated with GPIIb/IIIa as previously described (*39*). Briefly, GPIIb/IIIa (0.1 μg/well) was incubated overnight at 4°C in Buffer A (20mM Tris-HCl, pH 7.5, 150 mM NaCl, 1 mM $CaCl_2$, 1 mM $MgCl_2$, 1 mM $MnCl_2$). Following coating, plates were washed with Buffer B (50 mM Tris-HCl, pH 7.5 , 100mM NaCl, 1mM ea. $CaCl_2$, $MgCl_2$, and $MnCl_2$, 0.5% Tween 20) and wells blocked with Buffer B containing 35 mg/ml BSA (Fraction V, Sigma) for 2 hours at 30°C. Following blocking, wells were washed with Buffer B and phage added (1×10^8/well) to wells either in the absence or presence of increasing concentrations of soluble RGD peptide competitor (GRGDTP), diluted in Buffer B containing 1 mg/ ml BSA. As a phage background binding control, phage were allowed to incubate in wells coated with BSA, prepared and processed similarly. After one hour incubation at room temperature, wells were washed extensively with Buffer B, and phage eluted with 100 μl of 0.1 N HCl, pH 2.5 for 15 minutes with gentle rocking. An aliquot (20 μl) of the eluate was then removed and neutralized with 10 μl of 1 M Tris, pH 9.0. Starved *E. coli* K91Kan cells (30 μl) were added to neutralized phage eluates, incubated 15 minutes at room temperature, at which time LB containing 0.2 μg/ml tetracycline and 100ug /ml kanamycin was added, and the incubation continued with vigorous shaking for an additional 30 minutes at 37°C. Finally, 2.5 ml of LB containing 50 μg/ml tetracycline and 100 ug/ml kanamycin was added and cultures were placed, without shaking, at room temperature for overnight incubation. The following day, cultures were incubated at 37°C with shaking until the cultures became visibly turbid. At that time, 100 μl of 10% NaN_3 was added to the cells to arrest growth and their O.D. (optical density) read at 600 nm.

Modeling of CCSL miniprotein.

All 3-dimensional modeling of the CCSL miniprotein was carried out using the molecular modeling package SYBYL, version 6.1a (Tripos Associates, St. Louis, MO). The backbone template was derived from helices 1 and 3 of the asymmetric unit cell of a 29-residue synthetic peptide, coil-Ser (*40*). Sequence alignment of the CCSL miniprotein to these 2 helices was performed using GCG software (University of Wisconsin). Following sidechain replacement, three close contacts (<3Å) were manually corrected and the sidechain of Arg43 was rotated to introduce a potential interhelical salt bridge with Glu19. A conjugant gradient minimization was carried out for 500 iterations using the Kollman all-atom force field (*41*) with a distance dependent dielectric, $\varepsilon = 1$, and a non-bonded cutoff of 8Å. Following minimization, the RMS fit of the backbone atoms (C, Cα, N, O, Cβ) of the CCSL model to the coil-Ser template was 0.31Å.

Results

Construction and cloning of CCSL miniprotein on phage

Following PCR assembly of the CCSL/FLAG/Factor Xa prototype, cloning into the fUSE5 vector, and selection of phage clones expressing the FLAG immunoaffinity tag, several phage isolates were examined by DNA sequencing. One clone, nearly perfectly matched in sequence with the desired prototype, contained 3 missing bases within the 5' region of the CCSL DNA construct, changing the reading frame of the region encoding approximately the first 5 residues (data not shown). The proper reading frame, however, by virtue of the third point deletion, was restored such that the (distal) portion, including the FLAG tag, was deduced to be in-frame. These deletions were likely introduced by PCR during the assembly process. This clone was corrected by PCR mutagenesis and a corrected prototype clone verified by DNA sequencing, designated as CCπ.1 (Figure 3A). The deduced protein sequence of CCπ.1 is shown in Figure 3B.

A

    1   ggccgacgcg gccgttgccg ctctagaaag cgaggttagt gcccttgaga

   51   gcgaagtggc aagcctagaa agcgaggtcg cggcactcgg tcgaggcgat

 101   atgccattag ctgccgttaa gagtaagttg agtgccgtaa aggcaaaatt

 151   ggcttcggtc aagagcaagc tggcagcggt cgctgattac aaggacgatg

 201   acgatgccgc aatcgaggga agagcagcgg ccgctgtggc c

B

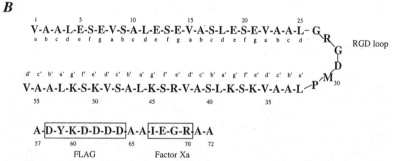

Figure 3. CCSL prototype clone sequencing. Panel **A** shows results of DNA sequencing of the CCπ.1 phage clone. **B.** Deduced protein sequence of the 72 residue CCSL-containing clone, CCπ.1. The positions of the heptad are designated a,b,c,d,e,f and g for residues 1-25 and a', b' c', d', e', f', g' for residues 32-56. The FLAG immunoaffinity tag and Factor Xa cleavage sites are denoted, as is the RGD-containing stem loop region.

Phage-displayed CCSL prototype expresses FLAG epitope

Initial evidence for the correctly-deduced amino acid sequence of CCπ.1 was obtained by an ELISA examining FLAG epitope expression. As shown in Figure 4, two CCSL phage clones, intermediate clone WO4 and prototype CCπ.1, bound to M2 anti-FLAG mAb-coated wells. Similar binding was observed with a control phage expressing the FLAG (DYKDDDD) (Miceli, R. M., unpublished data) peptide epitope. In contrast, a phage clone chosen at random from a library of decapeptide-expressing phage, R10, (AESTVHGPVE)(*37*) showed no binding to anti-FLAG-coated wells. All phage clones, however, demonstrated binding to the control anti-M13 phage antibody coated wells. These results clearly demonstrate that the FLAG epitope (DYKDDDD), expressed as the COOH-portion of the CCSL CCπ.1 prototype, is available on the phage surface for affinity interactions. In addition, it suggests that the CCSL miniprotein sequence is properly translated and expressed as a fusion with the pIII minor coat protein on the phage surface.

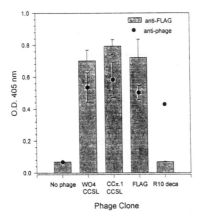

Figure 4. FLAG epitope expression on CCSL-containing phage. PEG-purified phage clones were added to either M2 anti-FLAG mAb-(*hatched bars*) or anti-M13 phage (*filled circles*) antibody-coated wells. Following incubation, wells were washed and incubated with anti-phage HRP-conjugate, washed again, and color-developed with ABTS (see *Methods* for details). Assay results are represented by an average O.D. 405 nm calculated from triplicate wells (+/- standard deviation).

Binding of RGD-containing CCSL phage to GPIIb/IIIa integrin receptor

To further demonstrate the expression of the CCSL prototype on phage and the availability of the loop region of the miniprotein to recognize its counter-receptor, we examined the ability of CCSL-phage containing the RGD motif to bind to integrin receptor GPIIb/IIIa. Phage containing the RGD sequence (CCπ.1 CCSL) demonstrated specific competition with soluble RGD peptide in binding to GPIIb/IIIa (Figure 5). In contrast, phage lacking the motif (FLAG,

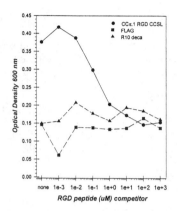

Figure 5. RGD-containing CCSL phage binding to integrin receptor GPIIb/IIIa. PEG-purified phage clones were added (1×10^8/well) to wells coated with GPIIb/IIIa, in the absence or presence of increasing concentrations of soluble competitor RGD peptide. Following incubation, wells were washed extensively and bound phage eluted with acid. Eluates containing the phage (tet[r]) were then used to infect 'starved' *E. coli,* and cells allowed to grow in tetracycline-containing medium. Points represent O.D.$_{600\ nm}$ measurements of the turbidity of *E. coli* cultures following the outgrowth period.

R10 deca), showed no effect of increased RGD peptide concentration for binding to GPIIb/IIIa-coated wells and demonstrated binding only slightly higher than BSA background controls, which were typically in the range of 0.05-0.10 O.D.$_{600}$ units. These competition binding data for RGD-containing CCSL miniproteins on phage agree well with a report by Healy et al. (*42*), who isolated and characterized hexameric RGD-containing peptide ligands for integrin GPIIb/IIIa from a phage display library. Further, these results validate that the CCSL is expressed on phage suitably to allow accessibility of the RGD peptide motif, encoded in the miniprotein stem loop, in binding to immobilized integrin receptor.

Modeling of the CCSL miniprotein phage prototype

In an effort to understand the potential usefulness of our CCSL phage prototype and aid in the design of peptide libraries expressed within its scaffolding, we modeled the CCSL helices based on homology to a synthetic peptide which forms a triple-stranded α-helical bundle in the crystal state, coil-Ser (*40*). Primary sequence alignment, based on residue heptad characterization and maximum homology, was performed for both CCSL helices with the antiparallel -strands of the coil-Ser bundle (NH_2-E$_g$W$_a$E$_b$A$_c$L$_d$E$_e$K$_f$K$_g$L$_a$-A$_b$A$_c$L$_d$E$_e$S$_f$K$_g$L$_a$Q$_b$A$_c$L$_d$E$_e$K$_f$K$_g$L$_a$E$_b$A$_c$L$_d$E$_e$H$_f$G$_g$-COOH) (*40*). These substitutions resulted in matches of 44 and 40% identity (56 and 60% similarity) between the CCSL helix and the coil-Ser sequence. Substitution of

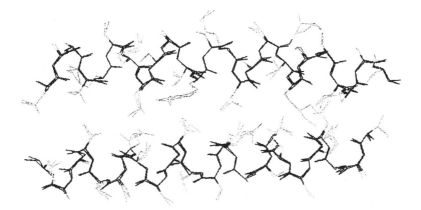

Figure 6. Superimposition of pre- and post-energy minimized CCSL structures. Models are based on superimposition of the phage prototype CCSL sequence to the coil-Ser peptide crystal structure described by Lovejoy et al. (*40*).

the CCSL residues into coil-Ser resulted in only minor discrepancies in the structure. Three close contacts (<3.0 Å), the results of β-branches introduced as the consequence of residue substitutions, were revealed upon inspection. These contacts were relieved by single bond rotation prior to energy minimization. In addition, the sidechain of Arg43 was rotated to better accommodate a potential intrahelical salt bridge between this residue and Glu19. As shown in Figure 6, an overlay of the pre- and post-minimization structures of the CCSL model shows only minor changes. In addition, the positions of the backbone atoms of the minimized model relative to the coil-Ser template are virtually unchanged (RMS deviation = 0.31Å). These results suggest that the predicted helical sections of the CCSL miniprotein can interact in a manner similar to that observed for the antiparallel arrangement of helices 1 and 3 in coil-Ser.

Discussion

We report here an advancement of the CCSL as a tool for α-helical and loop discovery and mimicry by its successful display on phage. Our earlier analyses using circular dichroism, size exclusion chromatography and sedimentation equilibrium measurements, have shown this simplified coiled coil sequence adopts a stable monomeric intramolecular antiparallel coiled coil stem loop in aqueous solution. (*26*). We now demonstrate phage CCSL prototype expression indirectly through a COOH-terminal immunoaffinity tag and more directly by the ability of the loop region to act as a ligand for integrin receptor GPIIb/IIIa. The latter marks a pivotal assay in establishing the prototype as a presentation scaffold for peptide library display. Characterization of affinity-enriched CCSL variants will also be potentially easier via the inclusion of a

Factor Xa cleavage site near the fusion of the construct to minor coat protein, pIII. Consequently, cleaved CCSL miniproteins, containing the FLAG immunoaffinity tag, can be used for binding assays without any steric hindrance mediated through attachment to the phage particle. Further, the CCSL/FLAG/Factor Xa construct has been engineered to allow incorporation of peptide libraries at strategic points within the construct. Using a PCR-generated cassette insert to randomize key residues within either loop or helix, we are currently generating CCSL-based constrained peptide libraries. One envisioned application of these may be their use in a two-stage selection scheme, similar to that used to delineate core residues and structural range within an epitope (37).

Homology modeling of our CCSL miniprotein to two antiparallel strands of the coil-Ser bundle described by Lovejoy et al. (40) has yielded the working structural model shown in Fig. 6. That there is sequence interchangeability between these two coiled coil structures is consistent with available data for known coiled coil sequence-structure relationships. Studies focusing on residue requirements for helical tendency and stability have concluded that these structures are very tolerant to sequence changes, as long as these changes adhere to the general nature of the hydrophobic and hydrophilic heptad repeat. The close RMS fit (0.31Å) between the pre- and post- energy minimized CCSL modeled on the antiparallel strands of the coil-Ser peptide crystal structure is supportive of the idea that the CCSL can adopt a very similar or identical structure. We feel this model will aid us in the future for conceptualizing the design of CCSL variants and libraries. However, a limitation of the current modeling is that it is based on coil-Ser, which contains two antiparallel coiled coil pairings in a three-stranded helical bundle and hence may not be fully representative of a dimer coiled coil (40). An alternative would be to employ the antiparallel dimer coiled coil motif available from the seryl tRNA synthetase structure (50). This structure will also accomodate the sequence of the CCSL miniprotein . Arguments could be made in favor of one model over the other. However, we believe it is premature to suggest that either model is currently a better structural representation than the other. For now, we take the model structure in Fig 6 as a reasonable working model.

A recent focus of many groups has been the advance of protein scaffolds for presentation and manipulation of constrained libraries of peptides. Several reports of engineered binding or recognition domains in *de novo* designed proteins have appeared. One study described the display of a conformationally homogenous combinatorial peptide library on phage (43) whose design is based on a previously characterized α-helical metal-binding motif (44). Aided by the linkage between metal coordination and the folding of this 'structure-induced' zinc-finger, metal-dependent epitopes to an antibody receptor were successfully identified. The experimental linkage of metal binding and folding may favor isolation of peptide motifs or sets of motifs that prove more useful in predicting pharmocophore structures. An interesting alternative approach in design of engineered specificity was recently described. An α-helical domain, derived from TFEB, a helix-loop-helix DNA binding protein, was inserted within the CDR3 region of an antibody Fab (45). The Fab-Ebox construct demonstrated binding specificity similar to native TFEB for DNA recognition sequence motifs. Another recent paper described the incorporation of T and B cell epitopes of the human immunodeficiency virus (HIV) into a *de novo* designed four-helix bundle (*46*). The synthetic protein vaccine was effective in eliciting both humoral and cellular immune responses to HIV-1. However, no structural information about this construct was presented.

We believe that α-helical arrays, and in particular coiled coils, are important scaffold candidates. First, they have been structurally well characterized. Coiled coil geometry and stoichiometry have been studied utilizing both 2D NMR (*9*) and high-resolution crystal structures (*14, 40*). These structural studies have significantly aided the understanding of the helical arrays and initiated an interest in these units as platforms for *de novo* protein design. Second, coiled coils are tolerant to significant primary sequence changes, while still maintaining their helical integrity. This has been reinforced by the introduction of metal-binding sites into several coiled coil designs (*47, 48*). In addition, a *de novo* designed CCSL containing sequence motifs from the A and D helices of human IL5 4-helix bundle cytokine, folds to adopt an α helical structure under aqueous conditions (*49*). Lastly, in the case of phage-expressed CCSL, this array can express both helical and loop surfaces and potentially be selected via affinity based on specific interactions mediated by these surface residues.

Acknowledgements

The authors are greatful to SB scientists Drs. Larry Helms and Ron Wetzel for help with the GPIIb/IIIa binding assay and to Joyce Mao, Rene Morris and Dr. Ganesh Sathe for support with oligonucleotide synthesis and DNA sequencing.

Literature Cited

1. Bazan, J.F. Science **1992**, *257*, 410-413.
2. Redfield, C.; Smith, L.J.; Boyd, J.; Lawrence, G.M.P.; Edwards, R.G.; Smith, R.A.G.; Dobson, C.M. Biochemistry, **1991**, *30*, 11029-11035.
3. de Vos, A.M.; Ultsch, M.; Kossiakoff, A.A. Science, **1992**, *255*, 306-312.
4. Kadesch, T. Immunology Today, **1992**, *13*, 31-36.
5. Kabat, E.A.; Wu, E.T.; Perry, H.M.; Gottesman, K.S.; Foeller, C. In *Proteins of Immunological Interest;* U.S. Department of Health and Human Service, NIH: Bethesda, MD, 1991.
6. Brown, J.E.; Klee, W.A. Biochemistry, **1971**, *10*, 470-476.
7. Bierzynski, A.; Kim, P.S.; Baldwin, R.L. Proc. Natl. Acad. Sci. U.S.A., **1982**, *79*, 2470-2474.
8. Hodges, R.S.; Semchuk, P.D.; Taneja, A.K.; Kay, C.M.; Parker, J.M.R.; Mant, C.T. Pept. Res., **1988**, *1*, 19-30.
9. Oas, T.G.; McIntosh, L.P.; O'Shea, E.K.; Dahlquist, F.W.; Kim, P.S . Biochemistry, **1990**, *29*, 2891-2894.
10. Cohen, C.; Parry, D.A.D. Prot. Struct. Funct. Genet., **1990**, *7*, 1-15.
11. Landschulz, W.H.; Johnson, P.F.; McKnight, S.L. Science, **1988**, *240*, 1759-1764.
12. O'Shea, E.K.; Rutkowski, R.; Kim, P. S. Cell, **1992**, *68*, 699-708.

13. Mo. J.; Holtzer, E.; Holtzer, A. Biopolymers, **1990**, *30*, 921-927
14. O'Shea, E.K.; Klemm, J.D.; Kim, P.S.; Alber, T. Science, **1991**, *254*, 539-544.
15. McLachlan, A.D.; Stewart, M. J. Mol. Biol., **1975**, *98*, 293-304.
16. Engel, M.; Williams, R.W.; Erickson, B.W. Biochemistry, **1991**, *30*, 3161-3169.
17. Graddis, T.J.; Myszka, D.G.; Chaiken, I.M. Biochemistry, **1993**, *32*, 12664-12670.
18. Rozzelle, J.E.; Wagner, D.S.; Tropsha, A.; Erickson, B.W. In *Peptides: Chemistry, Structure and Biology;* Hodges, R. S.; Smith, J. A., Eds.; Alberta University Press, Edmonton, 1993.
19. Richardson, J.S.; Richardson, D.C. In *Prediction of Protein Structure and Principles of Protein Conformation ;* Fasman G. D., Ed.; Plenum Press, NY, 1989; pp. 1-98.
20. Regan, L.; DeGrado, W.F.Science, **1988**, *241*, 976-978.
21. O'Neil, K.T.; Hoess, R.H.; DeGrado, W.F. Science, **1990**, *249*, 774-778.
22. Kamtekar, S.; Schiffer, J.M.; Xiong, H.; Babik, J.M.; Hecht, M.H. Science, **1993**, *262*, 16801685.
23. Lupas, A.; Van Dyke, M.; Stock, J. Science, **1991**, *252*,1162-1164.
24. Lau, S.Y.M.; Taneja, A.K.; Hodges, R.S. J. Biol. Chem., **1984**, *259*, 13253-13261.
25. Myszka, D.G.; Chaiken, I.M. In *Peptides: Structure, Chemistry, and Biology.* Hodges, R.S., Ed.; ESCOM, Leiden, 1993.
26. Myszka, D.G.; Chaiken, I.M. Biochemistry, **1994**, *33*, 2363-2372.
27. Scott, J.K.; Smith, G.P. Science, **1990**, *249*, 386-390.
28. Cwirla, S.E.; Peters, E.A.; Barrett, R.W.; Dower, W.J. Proc. Natl. Acad. Sci. USA, **1990**, *87*, 6378-6382.
29. Devlin, J.J.; Panaganiban, L.C.; Devlin, P.E. Science, **1990**, *249*, 404-406.
30. Lowman, H.B.; Bass, S.H.; Simpson, N.; Wells, J.A.. Biochemistry, **1991**, *30*, 10832-10838.
31. McCafferty, J.; Johnson, R.H.; Chiswell, D.J. Protein Eng., **1991**, *4*, 995-961.
32. Roberts, B.; Markland, W.; Ley, A.C.; Kent, R.B.; White, D.W.; Guterman, S.K.; Ladner, R.C. Proc. Natl Acad. Sci. U.S.A., **1992**, *89*, 2429-2433.
33. Hoogenboom, H.R.; Marks, J.D.; Griffiths, A.D.; Winter, G. Immunol. Rev., **1992**, *130*, 41-57.
34. Di Donato, A.; de Negris, M.; Russo, N.; Di Biase, S.; D'Alessio, G. Anal. Biochem., **1993**, *212*, 291-293.
35. Parmley, S.F.; Smith, G.P. Gene, **1988**, *73*, 305-318.
36. Horton, R.M.; Hunt, H.D.; Ho, S.N.; Pullen, J.K.; Pease, L.R. Gene, **1989**, *77*, 61-68.
37. Miceli, R.M.; DeGraaf, M.E.; Fischer, H.D. Journal of Immunological Meth., **1994**, *167*, 279-287.
38. Sambrook, J.; Fritsch, E.F.; Maniatis, T. *Molecular Cloning: A Laboratory Manual.*, Second Edition; Cold Spring Harbor Press, U.S.A., 1989.

39. Smith, J.W.; Ruggeri, Z.M.; Kunicki, T.J.; Cheresh, D.A. J.Biol. Chem., **1990**, *265*, 12267-12271.
40. Lovejoy, B.; Choe, S.; Cascio, D.; McRorie, D.K.; DeGrado, W.F.; Eisenberg, D. Science, **1993**, *259*, 1288-1293.
41. Weiner, S.J.; Kollman, P.A.; Nguyen, D.T.; Case D.A. J. Comp. Chem., **1986**, *7*, 230-253.
42. Healy, J.M.; Muratama, O.; Maeda, M.; Yoshino, K.; Sekiguchi, K.; Kikuchi, M. Biochemistry, **1995**, *34*, 3948-3955.
43. Bianchi, E.; Folgori, A.; Wallace, A.; Nicotra, M.; Acali, S.; Phalipon, A.; Barbato, G.; Bazzo, R.; Cortese, R.; Felici, F.; and Pessi, A. J. Mol. Biol. **1995**, *247*, 154-160.
44. Pavletich, N.P.; Pabo, C.O. Science, **1991**, *252*, 809-817.
45. McLane, K.E.; Burton, D.R.; Ghazal, P. Proc. Natl. Acad. Sci. U.S.A., **1995**, *92*, 5214-5218.
46. Eroshkin, A.M.; Karginova, E.A.; Gileva, I.P.; Lomakin, A.S.; Lebedev, L.R.; Kamiyinina, T.P.; Pereboev, A.V.; Ignat'ev, G.M. Protein Engineering, **1995**, *8*, 167-173.
47. Regan, L.; Clarke, N.D. Biochemistry, **1990**, *29*, 10878-10883.
48. Handel, T.M.; Williams, S.A.; DeGrado, W.F. Science, **1993**, *261*, 879-885.
49. Miceli, R.; Myszka, D.; Mao, J.; Sathe, G.; Chaiken, I. Drug Design and Discovery, **1996**, *in press*.
50. Cusack, S.; Berthet-Colominas, C.; Hartlein, M.; Nassar, N.; Leberman, R. Nature, **1990**, *347*, 249-255.

Automated Solid-Phase Synthesis

Chapter 17

Automation of High-Throughput Synthesis
Automated Laboratory Workstations Designed To Perform and Support Combinatorial Chemistry

James R. Harness

Bohdan Automation, Inc., 1500 McCormick Boulevard, Mundelein, IL 60060

A series of Workstations have been developed to perform both solution-phase and solid-phase organic synthetic reactions and to support related activities for the combinatorial chemistry laboratory. Workstations capable of performing resin dispensing, reagent preparation, chemical reaction preparation, product processing, and product aliquotting/archiving will be discussed. Information about sample tracking and software interfacing will be addressed in conjunction with specific Workstations.

Automated Laboratory Workstations provide a modular mechanism for automating a variety of sample handling processes. The Workstation approach offers the advantages of simplicity, throughput, reliability, and versatility and, therefore, is an extremely well suited automation strategy for supporting such a complex process as high throughput synthesis. A series of Workstations and related equipment capable of performing all aspects of organic synthesis will be discussed in this chapter.

For the purpose of this discussion, high throughput synthesis will be divided up into the series of functions shown in Figure 1. Beginning with reagent preparation and concluding with sample archiving covers most of the upstream and downstream support activities surrounding the synthesis process itself. These support activities are very important considerations to include in a high throughput synthesis discussion simply because these can become bottlenecks that limit the efficiency of the entire process. The Workstations to be described here will perform functions that include reagent preparation, reaction preparation, reaction cleavage/work-up, and product archive.

1054–7487/96/0188$15.00/0

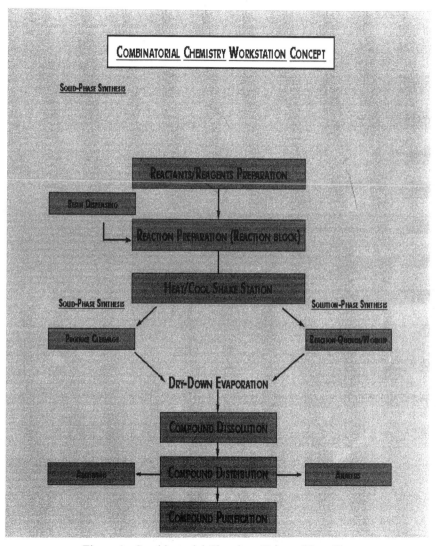

Figure 1. Combinatorial Chemistry Process Flow Chart.

Reagent Preparation

In order to simultaneously perform a series of organic synthetic reactions, a variety of different chemical reagents or reactants must be prepared. Depending on the nature and diversity of the reactions to be performed, this could involve preparing anywhere from a dozen or so reactants to up to a hundred or more. Reactants can also be prepared in advance and archived for later use, and this process is more efficiently accomplished by doing large batches.

The considerations that must be addressed in order to effectively design a Workstation for this purpose include the need for the following functions: bar-code reading and labeling, weighing, mixing, inert gas purging, and solvent dispensing.

Workstation Features. The Workstation layout shown in Figure 2 is a top view of a unit specifically designed to support reagent preparation. The unit contains a bar-code printer/labeler, bar-code reader (both hand-held and mounted), 3-place pan balance, vortex mixer, syringe drive with 6-way valve, three function cannula with wash station, and sufficient space for a reactants rack. The reactants containers shown on this machine are 40 mL scintillation vials that receive a septum cap. The cannula is a septum-piercing, venting, gas purging cannula that is capable of venting and purging these vials with inert gas while simultaneously adding or removing solvent. The Workstation contains one pair of grippers that are used to move the reactant vials to the balance, mixer, printer/labeler, and bar-code reader.

Sequence of Operation. The equipment functions in two basic modes. The first process involves operator assisted dispensing of reactants into the reactants vials. The second operation fully automated dilution and dissolution of the dispensed reactants.

Reactant Dispensing. Operator assistance is required for the reactants dispensing operation in order to be able to accommodate the wide variety of materials physical chemical properties. The Workstation is used to perform all container movements and, therefore, automatically creates a tracking record for all activities performed. The specific sequence of operations that the Workstation performs is shown below:

1) Zero the balance, move container #1 (uncapped) to the balance, determine the tare weight, and re-zero the balance.
2) The operator manually scans the bar-code label on the first reagent stock bottle, adds the specified amount of material to container #1 and signals the Workstation (computer keystroke) that the operation is complete.
3) The Workstation determines the container gross weight.
4) The operator manually places a cap on the vial and places the vial in the bar-code reader nest.
5) The Workstation returns the vial to its original location in the reactants rack.
6) Steps 1-5 are repeated until all reactants have been added to vials.

Dilution and Dissolution. This part of the protocol does not require operator assistance. Information regarding dilution concentrations (molar, etc.) must be input to the unit, but this can be accomplished by file transfer from a LIMS or other network system. At this point, the Workstation performs the following protocol:

1) Purge reactant container #1 with inert gas and add the appropriate volume of the appropriate solvent.
2) Vortex mix the contents of container #1 and bar-code label the vial with the appropriate information.
3) Return container #1 to its original location in the reactants rack.
4) Repeat Steps 1-2 until all vials are prepared.

The reactants in this run are now ready for use in a synthesis run or can be archived for later use. Archived vials can be brought back to this Workstation for bar-code reading prior to being placed on the reaction preparation Workstation. Reading the bar-codes on the Workstation creates a file of information regarding the order of the vials in the rack and this information can be transferred to the next unit.

Reaction Preparation

The reaction vessels used to contain the reaction materials and the reaction block that will house the vessels will not be discussed in detail in this chapter. The containers are typically glass and the block is capable of being heated and cooled over the range from -40°C to 150°C. This device can be sealed to hold pressures up to 2 atmospheres or can support reflux-requiring procedures. The incubations are performed on a separate shaking station so that the Workstation can be available for setting up additional reactions.

Setting up a series of organic reactions to be run simultaneously in a single reaction block involves the ability to perform numerous liquid handling functions while maintaining an inert and, possibly, cold environment. The ability to mix reactants immediately after addition to the reaction vessels is also critical to prevent uncontrolled and undesired side reactions from occurring.

Workstation Features. Figure 3 shows a top view of a Workstation specifically designed to perform reaction set-up or mapping. The unit has space for 3 different rack types (reactants, common reagents and the reaction block), a three function cannula attached to a syringe drive and 4-way valve, a wash station for the cannula, and a nest for a stock reagent. The reaction block sets in an argon purged box so that an inert environment is maintained in and around the reaction vessels. The block also sets on an orbital shaker platform that provides for mixing at critical stages during the reactant/reagent addition steps. The mapping cannula is a septum piercing, venting, argon purging cannula that provides a mechanism for retaining the inert environment with the reaction vessels and the reactants/reagents containers while moving solutions around. The resin washing/solvent dispensing cannula is a multi-channnel dispenser that is used for addition of solvent, but does not do reagent or reactant transfers. All containers are septum capped for environmental control. The reactants containers are

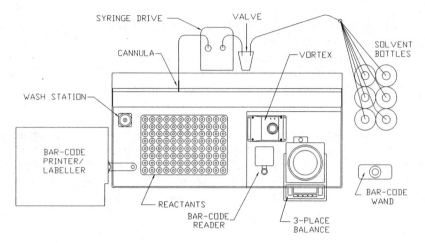

Figure 2. Automated Reagent Preparation Workstation.

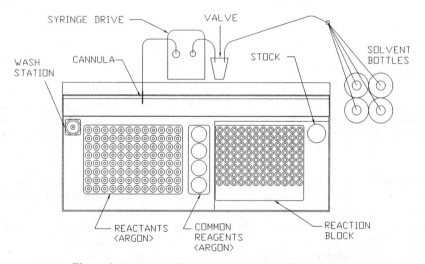

Figure 3. Automated Reaction Preparation Workstation.

40 mL scintillation vials (from the reagent preparation Workstation) while the common reagents and stock bottles can be 50, 100, 250, or 500 mL containers.

Sequence of Operation. The protocols employed for setting solid-phase reactions vs solution-phase reactions are different at a number of steps. The sequence of operation presented below will be a general protocol that will point out these differences at specific steps. Either process may require that reactants be pre-mixed immediately prior to addition to the reaction vessel. This can be accomplished on the Workstation, but will not be highlighted in the general procedure. The addition of solid reactants to the reactions vessels during the course of a reaction set-up can be handled, but involves operator assistance and cannot be accomplished solely by the Workstation.

The information specifying the details the reaction mapping can be input through the Workstation using the "sequence building" software. This information can also be provided to the Workstation in the form of a text file from a LIMS or other network system. As with the reagent preparation Workstation, tracking of items involved in the process can be accomplished by scanning bar-code labels from all racks and/or reagent bottles used in the run.

1) For a solid-phase reaction, wash the resin with solvents. For solution-phase, dispense the stock reagent (building block) into all reaction vessels.
2) While purging reactant container #1 with inert gas, aspirate the selected volume of material and dispense into the appropriate reaction container(s).
3) Repeat Step #2 for all reactant transfers and additions defined in the map.
4) While purging common reagent container #1 with inert gas, aspirate the selected volume of material and dispense into the appropriate reaction container(s).
5) Repeat Step #4 for all transfers and additions defined in the map.
6) Mix the contents of the vessels in the reaction block.

At this point in the process, the operator would place the appropriate top on the reaction block and move the block to a separate shaker station. This station can heat the block under pressure (sealed containers) or reflux the contents of the vessels by heating one layer and cooling a second layer in the block (vented top). The incubation of the reactions continues with mixing for the specified time. Solid-phase reactions may return to the reactant preparation Workstation for a wash, follow up round of reactants/reagents and return to the shaker for incubation.. At the completion of the reactions, the block would be transferred to the work-up/cleavage Workstation.

Work-up/Cleavage

The post-synthesis activities involved in solid-phase reactions differ substantially from those used for solution-phase reactions. Both types of processes can be readily automated using the Workstation concept and can even be built into the same unit. Solution-phase reactions typically involve addition of a quench solution, such as a dilute acid or base, to stop the reaction and begin the work-up process. Quenching may be followed by a series of liquid/liquid extractions to remove reactants, effect a solvent exchange, or extract product. After processing, product is transferred into containers that permit easy dry-down and dissolution into solvents that support downstream activities. These functions can be and have been performed on automated devices, such as Workstations.

Solid-phase synthesis reactions require removal of the product from the solid support that was employed during the synthesis. The cleavage of product from the support typically requires the use of a strong acid. A common acid used for this purpose is trifluoroacetic acid (TFA). The collection of the cleave product requires a separation of the liquid from the solid and can easily be performed with pressure or vacuum across a membrane. Substantial care must be taken in the handling of the TFA, because of the corrosive nature of this compound.

For this discussion, the Workstation features described will detail the components required for both solution-phase and solid-phase reaction work-ups. However, the sequence of operation described will only handle post-synthesis activities for a solid-phase reaction using TFA as the cleavage reagent.

Workstation Features. The Workstation layout shown in Figure 4 is a top view of a unit specifically designed to perform solution-phase work-up. The unit contains a hand-held bar-code reader (not shown), syringe drive with 2-way valve, syringe drive for aqueous solutions, individual wash stations to separate organic from aqueous waste, and space for the reaction block and a set of dry-down tubes. The reaction block is setting on an orbital shaker to accommodate mixing requirements of the process. For solid-phase reaction support, the Workstation would have the dry-down tubes under the block for collection directly into the tubes and would have a space on the floor of the unit for well vented cleavage solution container. Solid-phase reactions would also require that the block be contained in an inert environment similar in nature to the set-up on the reaction preparation Workstation. The syringe drive dedicated for aqueous solutions on the solution-phase unit would be used for cleavage solution on the solid-phase unit. Washing the resins would again be handled with a multi-channel cannula while cleavage solution dispensing would be performed with a single channel cannula.

Sequence of Operation. Solution-phase work-up essentially is a series of liquid/liquid extractions with one phase being removed and discarded or placed in the dry-down tubes for further processing. The dry-down tubes are arranged in 4 racks that match the outer dimensions of a microplate. This provides a direct mechanism for transferring these tubes to a centrifugal evaporator for dry-down without having to move tubes from rack to rack. This same consideration applies to solid-phase reaction products.

The reaction block environment needs to remain inert until the reactants from the last reaction process have been removed to waste. This step is handled by the Workstation and then the following protocol is performed:

1) Wash the resin by adding the selected solvent, mixing to disperse the resin in the solvent, and aspirating the wash solvent to waste.
2) Manually the operator will place the collection rack under the block.

3) Add cleavage solution and mix.

4) Aspirate the cleavage solution into the dry-down tubes in the collection rack.

5) Perform a solvent wash, if selected, and add this material to the collected cleavage solution.

At this point, the collected product is ready for dry-down as described above. Reaction vessels can be cleaned and readied for another run.

Aliquotting/Archive

Dried-down product has to be prepared for chemical analysis, biological analysis, shipment, and/or long term storage. The solutions used to dissolve the products in order to remove a uniform aliquot vary for the different analyses to be performed. Biological analysis requires a solvent that is compatible with the biological materials used in the testing. Typically, dimethyl sulfoxide (DMSO) is used because this solvent has good solvating properties and limited denaturing concerns at low concentrations. On the other hand, this is poor choice of solvents for chemical analysis because of detection interferences. Different containers may also be required for the different purposes. Biological testing is often performed in 96-well plate formats, whereas analytical testing probably would prefer using a GC or LC glass vial. In addition, archiving is addressed best with glass, screw-cap containers that are bar-code labeled for positive identification. Therefore, the equipment must be able to accommodate a variety of considerations.

Workstation Features. The Workstation shown in Figure 5 shows only 2 rack types that are handled by the unit. These 2 rack types are for holding the dry-down tubes and microplates. Other rack types that can be placed on the Workstation include GC vial racks; screw cap, 1-dram vial racks; freezer boxes for 1-dram vials; and deep well plates. The unit also includes a feeder bowl for presenting caps to the capping station, bar-code printer/labeler, syringe drive with 2-way valve, and a pipetting cannula with wash station.

Sequence of Operation. This Workstation has a variety of protocols that can be performed with different combinations of racks. Only 2 of these protocols will be discussed in this section. These protocols will involve transferring an aliquot of product to a microplate for biological analysis and putting the remainder of the material in a 1-dram vial for long term storage.

Aliquotting. The rack combination shown in Figure 5 is used to support a process that includes dissolving the dried-down product in DMSO and aliquotting a portion of the material into a microplate for assay. The product can be dissolved to a fixed volume or dissolved to concentration. The latter requires information about the quantity of material in each tube. This is easily obtained by weighing these tubes before product is collected and, then, weighing again after the dry-down step is complete. This also permits calculation of product yield. The weighing process can

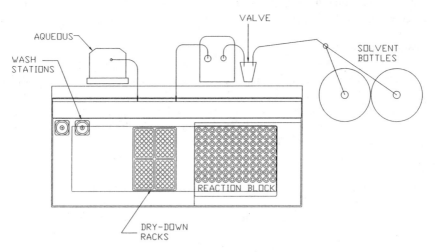

Figure 4. Automated Product Work-up/Cleavage Workstation.

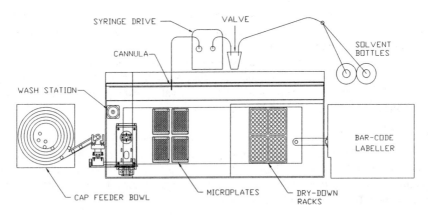

Figure 5. Automated Product Aliquotting/Archiving Workstation

be performed on a weighing Workstation and the results can be transferred via file transfer to a database and/or this Workstation. The product yields along with the molecular weights of the compounds provides the Workstation with sufficient information to permit dissolving materials to molar concentrations. The protocol followed to accomplish this is shown below:

1) Add DMSO to each tube to a final concentration of 1 mM.
2) Mix the tubes to effect complete dissolution.
3) Transfer an aliquot of each dissolved product to a well in the microplate and wash the cannula between each transfer.
4) Repeat Step #4 until all samples are transferred.

Archiving. The residual product that is in the dry-down tubes from the process above can now be put into 1-dram vials for long term storage. The 1-dram vials can be provided to the Workstation in racks that match the microplate footprint and the products can actually be dried down prior to capping and labeling. The process would be handled as listed below:

1) Transfer the contents of the dry-down tubes to the corresponding vials.
2) Manually, the operator dries down the materials in the vials in a centrifugal evaporator or similar device.
3) The Workstation caps vial #1, prints a bar-code label, applies the label to the vial, and places the vial in a freezer box.
4) Step #3 is repeated until all vials are capped, labeled and placed into storage boxes.

Summary

The process of high throughput synthesis can be fully automated. The Workstation approach to automating this procedure offers the general advantages of the Workstation concept along with the support of intermittent operator intervention. A procedure as complex and variable as organic synthesis is best monitored by chemist review at critical points in the process. The automation equipment and support devices used in the process should be specifically designed to require operator review at specific steps and to permit easy monitoring and interface by the attending chemist. Equipment-supported monitoring of the process is also an essential part of the overall considerations and should include liquid level sensing to avoid overflowing critical containers, such as, reaction vessels; pressure sensing in the liquid handling lines for detection of plugged cannulas from particulate or other non-pipettable materials (tars, etc.); and solvent vapor detection to monitor the integrity of container seals. Problems sensed by the equipment should be deferred to operator interpretation and support in the resolution.

The general design of the individual Workstations and the functions performed by each station should be tailored to match the processes being performed. This includes both the individual chemistries being performed and the containers/equipment used to perform the chemistries. Consideration must also be given to the physical plant (benches, hoods, etc.) that are available or can be accessed for performance of the process. The logic of trying to retrofit chemistries to different equipment and devices does not work very well for a procedure as complex as organic synthesis. Therefore, the automation should be built around and specific for the process being run manually. This means starting with the reaction container and reaction block and automating this portion first. Then, the upstream and downstream activities can be more exactly addressed.

Chapter 18

Boosting the Productivity of Medicinal Chemistry Through Automation Tools

Novel Technological Developments Enable a Wide Range of Automated Synthetic Procedures

Owen Gooding, Paul D. Hoeprich, Jr., Jeff W. Labadie,
John A. Porco, Jr., Paul van Eikeren[1], and Peter Wright

Argonaut Technologies, Inc., 887 Industrial Road, Suite G,
San Carlos, CA 94070

The authors describe novel approaches to automating solid phase organic chemistry through the use of an instrument, the *Nautilus*, and a resin, *ArgoGel*. The *Nautilus* synthesizer automates organic reactions through the use of a computer controlled fluid delivery system which is composed of glass and polytetrafluoroethylene and which is isolated from the outside atmosphere. *ArgoGel* resin is a novel polyethylene glycol grafted polystyrene polymer that provides higher loading and higher acid stability than existing gel phase resins.

A convergence of technological innovations and pressing needs within the pharmaceutical industry have combined to enable the development of tools which greatly enhance the productivity of medicinal chemists. Beginning with Merrifield's (1) revolutionary advances in peptide synthesis, chemists have long realized that it is possible to automate the synthesis of at least some types of organic molecules. Until recently, researchers focused on the automated synthesis of biopolymers not only because such syntheses entailed highly repetitive steps, but also because the relevant synthetic methodologies required only a narrow range of reactions, reagents and physical conditions. In the past, manual synthesis, not automation, was apropos for making new pharmacophores, since such molecules are associated with a broad universe of structural motifs and synthetic pathways. If a drug company wanted to make more compounds, it simply increased its research budget and hired more chemists.

Today, pharmaceutical price caps and billion dollar research and development budgets preclude increased headcount. Nonetheless, increased competition and declining drug exclusivity periods demand that medicinal chemistry departments increase the rate at which they produce new chemical entities. The problem is exacerbated by developments in biology, such as molecular biology and high-throughput-screening, which enable biology departments to screen entire archived compound libraries of major pharmaceutical libraries in mere months. The only plausible solution to the current bottleneck is a set of technologies which enable

---

[1]Corresponding author

chemists to automate the procedures that they currently perform in round bottom flasks.

Developments in the automation of organic synthesis have come about from technological advances in two principal areas. One, advances in fluid handling, robots, materials, electronics and software now make it possible to manufacture instruments that manipulate fluids consistent with the rigorous demands of organic chemistry. Such advances are build upon the elegant research of pioneering chemists who designed automated solution-phase synthesizers (2,3). And two, advances in solid phase synthesis, illustrated in Figure 1, in which a template molecule is immobilized onto an insoluble polymer bead and elaborated by multi-step treatment with reagents. Large excesses of reagents can be used to drive each reaction step to completion because spent reagents can be readily removed by repeated washing of the polymer bead with solvents.

Impressive advances in the repertoire of reactions that are useful in solid phase (4) synthesis are now enabling instrument manufacturers to design general-purpose, solid-phase, organic synthesizers. Though solution phase chemistry will always be at least partially required for synthesis, solid phase organic chemistry (and combinations of solid and liquid phase chemistry) has opened a variety of automation concepts.

Instruments for Organic Synthesis.

Instruments and resins must satisfy a number of requirements to be generally useful to medicinal chemists. In both cases, such products must give the researcher the ability to use the broadest possible range of reactions, otherwise, he will be obliged to develop new synthetic methodologies rather than using well characterized transformations. For most medicinal chemists, the goal is a new pharmacophore not a new reaction. Ideally, instruments should emulate and augment ordinary synthetic apparatuses. For most chemists, this means glass/polytetrafluoroethylene (PTFE) materials, the ability to heat and cool, an inert atmosphere, good mixing and visible reactions. Even though these requisite properties would seem obvious, previous generations of automation tools have fallen short in many of these areas thereby forcing researchers to spend time developing makeshift solutions.

In designing our instrumentation, our engineers worked closely with a number of prominent medicinal chemists to understand not only what features and specifications were desired, but also how the products could be constructed for optimal use in the laboratory. In designing our instrument, called the *Nautilus*, to meet the rigorous physical requirements of the laboratory, we had to engineer a number of technological innovations.   Since air and water (not to mention cross contamination) are an anathema to many organic reactions, it became clear that any robust automated synthesizer needed to be closed to the outside atmosphere. This is much easier to accomplish using a closed, fluid delivery system (similar to that used in DNA and peptide synthesis) than using an open gantry robot. Closed systems, however, add their own complexity. To handle the range of reagents used in ordinary glassware, we had to develop a system where all wetted surfaces are either glass or PTFE. The most difficult component was the instrument's automated "stopcocks" or valves. The valves were designed around a block of micromachined glass sealed with a PTFE membrane; the membrane functions as a barrier between the electro-mechanical actuators that open and close the valves and the reagents contained therein.

To complete the paradigm of an automated set of glassware, we developed an array of devices that, combined with our automatic valves, enables the introduction of precise aliquots of reagents into observable glass reaction vessels. Disposable glass "test tubes" take the place of round bottom flasks, while PTFE frits take the place of Buchner funnels in separating resins from reactants and solvents. The temperature of the vessels is regulated by a combination of chilled inert gas and a computerized heating mantle. This scheme enables one to independently control the temperature of each reactor. The geometry of the heating system was designed such that the inside of the reactor is visible at all times – an important requirement in reaction monitoring. The "stir bar" is replaced by a rocking agitator that inverts the reaction tubes, generates good turbulent mixing and eliminates physical contact that could damage resins. The specifications of the *Nautilus* instrument are summarized in Table I.

Table I. *Nautilus* Synthesizer Specifications

Fluid Delivery	Environment and Control
Glass/PTFE wetted surfaces	Graphical user interface
24 reaction vessel positions	Microsoft Access database
Glass reaction vessels, 8ml and 15ml	Programmable reaction procedures
112 reagent and solvent positions	Visible reaction chambers
Minimum reagent addition 200 μl	Rocking agitation
Maximum reagent addition 5.8 ml	-40 to 150 °C temperature control
3 segregated waste containers	
Fraction collector	

The physical embodiment of the synthesizer must be complemented with software that translates and automates benchtop operations, such as dropwise addition of reagent, with a minimum amount of computer expertise. Further, the software must be able to import data from corporate databases to minimize typed data entry. These requirements, coupled with the need to manage the physical operations of the machine, demand an elegant software solution that provides both a simple standardized user interface as well as a customizable interface to corporate databases. Our solution was to couple a programmable, proprietary graphical user interface to the instrument via an industry standard SQL database, Microsoft Access. This architecture enabled us to simplify the user interface such that the average user needs only a superficial understanding of the workings of the machine, while the more technically inclined innovator can program the instrument at a more detailed level. Finally, this approach enables corporate information systems personnel direct access to the database using the many available developer tools for Access.

Synthetic Chemistry on the *Nautilus*. To prove the capabilities of the synthesizer, we have run a broad variety of reactive solutions through the system to conduct several different multistep synthesis. Some of the more reactive species used on the synthesizer include: n-butyl lithium, lithium diethylamide (LDA), and sodium methoxide. Multistep reactions conducted on the machine include an enolate formation/alkylation (5) and a biaryl synthesis via a Suzuki coupling reaction (6). These reactions demonstrated both the materials compatibility of the *Nautilus*'s fluid delivery system and the ability of the instrument to maintain an inert environment at both high (90 °C) and low (0 °C) temperatures. The biaryl synthesis is detailed below and details of of the other procedures are available on request.

Biaryl synthesis via a Suzuki coupling. The "Suzuki" reaction has been frequently employed in solid phase organic synthesis and was judged to be representative of palladium-catalyzed carbon-carbon bond forming reactions. An example of biaryl formation from the literature, illustrated in Figure 2, was chosen as starting point for a series of automated organic synthesis experiments. For these experiments, the reaction solvent (DME) utilized in the literature example (6) was

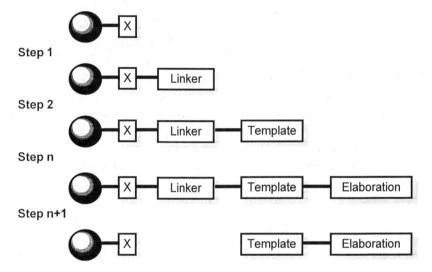

Figure 1. General scheme for solid phase organic synthesis

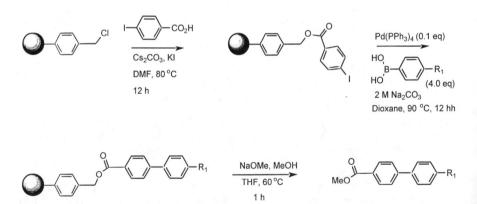

Figure 2. Biaryl synthesis via a Suzuki coupling.

changed to dioxane to facilitate mixing on the instrument. A solution of tetrakis-triphenylphosphine-palladium(0) in dioxane was prepared for catalyst deliveries. This stock solution could be left in a reagent bottle position on the instrument for several hours without precipitation of palladium black. Automated biaryl couplings were performed by consecutive additions of dioxane, Pd(0) solution, boronic acid in dioxane, and 2 M aqueous sodium carbonate solutions. After 8-12 h of heating at 90 °C, the resins were subjected to a washing cycle, including washes with 0.5% solution of diethyldithiocarbamic acid sodium salt/DIEA in DMF to facilitate removal of residual and precipitated palladium salts. Cleavage of biaryl products was achieved by saponification (NaOMe, THF/MeOH, 60 °C, 1 h). In general yields and purities of biaryl products, summarized in Table 2, were comparable to experiments performed manually in-house and described in the literature (6). Current work is focused on the demonstration of other palladium-catalyzed cross coupling reactions and chemistries that require either strict temperature control or inert environments.

Table II – Biaryl synthesis yield and purity.

R_1	Yield (%)	Purity (%)†
OMe	91	96.4
F	89	88.6

†Purity measured by HPLC: C_{18}, 10-100% acetonitrile, 215 nm

Resins for Organic Synthesis.

To make new pharmacophores via automated solid phase synthesis, chemists need resins that enable them to work with a majority of the universe of synthetic chemistry reactions. Ideal resins have characteristics that include high capacity, inert and stable to the conditions of the reaction, uniform bead size, low levels of leachable impurities, and high compatibility with a range of solvents. Most resins have intrinsic limitations. For example, polystyrene resins preclude exposure to strong electrophiles that lead to aromatic substitutions. Polyethylene resins generally exhibit extremely low capacities. To provide a complete "toolbox" of resins for solid phase organic synthesis, chemists need a variety of resin platforms that combine various polymers, tethers, linkers and functional groups.

Structure and Properties of *ArgoGel*. As a first step in providing a platform of resins, we developed a novel resin bead consisting a modified polystyrene backbone lightly (1-2 %) cross-linked with divinyl benzene that has been grafted with polyethylene glycol (PEG). These products, named *ArgoGels*, are described by the general structure shown in Figure 3 , where n is in the range of 30 to 40 units. *ArgoGel* is manufactured in alcohol (*ArgoGel–OH*®, X = OH) and amine (*ArgoGel–NH2*®, X = NH₂) forms. The resin was designed as an improvement of existing graft copolymers (8,9), which, when swollen by solvent, confer a solution-like environment to solid phase organic synthesis. The goal was to overcome some of the limitations of existing PEG resins, including, low capacity, lability of the benzylic ether-PEG linkage under strongly acidic conditions, and elution of interstitial PEG chains that are physically trapped in the resin, but which disentangle and co-elute with the cleavage of end products.

ArgoGel is a white to off-white solid composed of spherical beads approximately 150 μm in diameter. In the dry state, *ArgoGel* flows freely–a property that makes it easy to dispense samples into reaction vessels in an automated, reproducible, and accurate

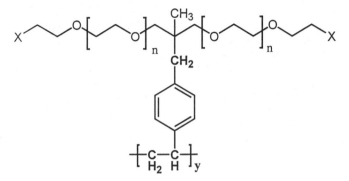

Figure 3. Structure of ArgoGel

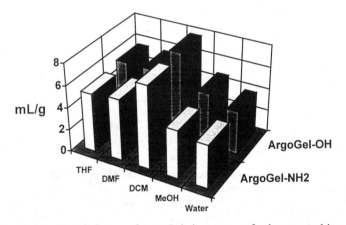

Figure 4. Swelling volumes of ArgoGels in a range of solvents used in solid phase organic synthesis

manner. *ArgoGel* particles exhibit a relatively narrow particle size distribution (150 ± 25 μm) with a very small percentage of fines (particles less than 100 μm comprise less than 2% of the sample). *ArgoGel* is a crystalline material exhibiting a sharp endotherm as measured by DSC (differential scanning calorimetry). The endotherm is associated with a phase transition in which the polyethylene glycol graft undergoes a change from the solid to the fluid state.

ArgoGel swells strongly in the presence of a range of solvents useful in solid-phase organic synthesis. Representative volumes for *ArgoGel–OH* and *ArgoGel–NH₂*, normalized for the weight (mL/g), in as set of solvents that are typically used in washing operation during solid-phase organic synthesis are shown in Figure 4. The high swelling volumes demonstrates that the polyethylene glycol graft is highly solubilized and that the behavior of the molecules bound to the support approximates molecules free in solution. Such behavior ensures solution-like diffusion rates resulting in rapid reactions and purification by washing.

ArgoGels **Exhibit High Capacities.** *ArgoGel–OH* and *ArgoGel-NH₂* exhibits capacities as high as 0.5 mmol per gram. Capacities are determined by measuring the FMOC capacity of FMOC-glycine coupled to the resin via a validated protocol. Higher resin capacities offer improved efficiency in solid-phase organic synthesis because they result in the production of larger quantities of synthetic product for the same-sized reaction vessel and same reaction time.

ArgoGel **Resins Contain Lower Impurity Levels.** *ArgoGel–OH* contains lower levels of leachable impurities than comparable products. For example, *ArgoGel–OH* incubated with trifluoroacetic acid/water (95/5; v/v) for 4 hours at room temperature typically releases only about 0.5 wt % impurities.. By comparison, Tentagel S OH® (Rapp Polymere) incubated under the same conditions releases 1 to 1.5 wt% impurities, principally polyethylene glycol. When normalized for the level of loading, *ArgoGel-OH* exhibits about a seven-fold reduction in the release of impurities relative to Tentagel S OH®. Thus, *ArgoGel* when used in solid-phase synthesis provides the user with products that exhibit significantly higher chemical purity.

Summary.

The instrumentation and resins discussed in this paper represent only the beginnings of a new industry centered around serving the automation needs of synthetic chemists. There is a need for many new devices and chemical products that can ease the tedious manual steps involved in organic synthetic procedures. Far from being a threat to organic chemists, these tools will enable chemists to focus on designing molecules, developing synthetic pathways and analyzing the mode of action of new pharmacophores. Based on its ability to satisfy the physical conditions and material compatibility requirements of solid phase organic synthesis, the *Nautilus* will provide medicinal chemists with a faster route to developing new solid phase organic synthesis methodologies, an automated facility to optimize lead compounds, and a ready means of re-synthesizing leads. Still, there is much room for innovative technology which can miniaturize synthesizers, increase the numbers of compounds synthesized and add further synthetic capabilities. If technological developments continue at their current pace, automated synthesizers may become as ubiquitous as HPLC instruments.

In a similar fashion, the range and capabilities of solid phase resins will expand with the demand for automated solid phase organic synthesis and combinatorial chemistry. *ArgoGel* is a first step in this direction with its high loading and chemical stability. Later resins will undoubtedly enable even higher yields and greater product purity– properties which will drive miniaturization and cost reduction. Furthermore, the development of polymers that are more inert and the creation of so-called "traceless" linkers (which leave no trace of the attachment point) will enable chemists to automate the synthesis of an ever broader range of pharmacophores. In total, these developments will drive not only the instrumentation industry, but also, and more importantly, a dramatic increase in the productivity of medicinal chemistry.

Acknowledgments.

We thank Argonaut's engineering team for their diligence and perspicacity in designing the *Nautilus* instrument, and we thank the members of our Product Development Consortium for their providing their requirements, suggestions and critiques.

References Cited.

1. Merrifield, R.B. *J. Am. Chem. Soc.* **1963,** *85,* 2149-2154.
2. Kramer, G.W.; Fuchs, P. *Byte* **1986,** *11,* 263-284.
3. Hayashi, N. et al. *J. Auto Chem* **1989,** *11,* 212-220.
4. Thompson, L.A.; Ellman, J.A. *Chem. Rev.* **1996,** *96,* 555-600.
5. Backes, B.J.; Ellman, J.A. *J. Am. Chem. Soc.* **1994,** *116,* 11171-11172.
6. Frenette, R.; Friesen, R.W. *Tetrahedron Lett.* **1994,** *35,* 9177-9180.
7. For a recent review on the Suzuki cross-soupling, see: Miyaura, N.; Suzuki, A. *Chem. Rev.* **1995,** *95,* 2457-2483
8. Bayer, E.; Rapp, W. In *Chemistry of Peptides and Proteins*; Voelter, W., Bayer, E., Ovchinnikov, Y.A., Ivanov, V.T. Eds; Proceedings of the Fifth USSR-FRG Symposium on Chemistry of Peptides and Proteins; Walter de Gruyter: Berlin, 1986, Vol. 3; pp 3-8.
9. Zalipsky, S.; Chang, J.L.; Alberico, F.; Barany, G. *Reactive Polymers* **1994,** *22,* 243-258.

Chapter 19

A Modular System for Combinatorial and Automated Synthesis

S. H. DeWitt, B. R. Bear, J. S. Brussolo, M. J. Duffield, E. M. Hogan,
C. E. Kibbey, A. A. MacDonald, D. G. Nickell, R. L. Rhoton,
and G. A. Robertson

Diversomer Technologies, Inc. and Parke-Davis Pharmaceutical
Research, 2800 Plymouth Road, Ann Arbor, MI 48105

DIVERSOMER technology is an enabling technology for the
implementation of combinatorial and automated synthesis. The
features of this technology include a modular design and open
architecture for the parallel synthesis of 8 or 40 compounds at one
time. This technology is amenable to both solution and solid phase
chemistries over a broad range of reaction conditions employing
manual and automated methods. Furthermore, the quantity and
quality of the compounds generated employing DIVERSOMER
systems allows for evaluation in multiple assays, quantitative
Structure Activity Relationship (SAR) data, and compound
conservation efforts.

Combinatorial and automated synthesis enables more efficient and productive
chemical discovery efforts (1-3). While the generation of combinatorial mixtures
provides large numbers of compounds at once, the high-throughput synthesis of
individual compounds provides smaller numbers of high quality compounds in
larger quantities. However, both approaches are valuable and complementary for
lead generation and lead optimization efforts.

The automation of synthetic processes has been implemented by a
relatively small number of dedicated efforts (4). The advent of combinatorial
chemistry has intensified interest in the application of automation to synthetic
chemistry. The decision to execute syntheses by manual or automated means is
often dictated by the necessary sample throughput and the compatibility of the
chemistry with existing instrumentation. In fact, manual methods are still the
optimum choice for several combinatorial chemistry approaches. For example, the
combinatorial synthesis of a single mixture containing 100,000 compounds is best

1054–7487/96/0207$15.00/0

achieved manually in a single reactor. Furthermore, the selection of solution phase or solid phase chemistry methods and the need for isolation and purification protocols influences the implementation of automation. Representative applications for automated synthesis include peptide synthesis, reaction optimization, traditional organic synthesis, and combinatorial chemistry (4).

Automated systems are designed to process operations in a serial or parallel fashion. A system that processes samples in parallel performs the same operation on a plurality of samples before proceeding to the next operation. In contrast, a serial processing system performs all of the operations on one sample before proceeding to the next sample. Serial processing systems often require scheduling software to maximize the use of the system. Without scheduling, for example, a serial processing instrument will wait idly during a single reaction that requires 24 hours. At this rate, less than 30 reactions could be achieved in one month. Parallel processing significantly impacts the throughput and efficiency of automated synthesis systems. For example, the protocol described above could be achieved in one 24 hour period.

The architecture of an automated synthesis system can be open or closed, modular or integrated. An open and modular system enables the use individual components with or without other components of system. This flexible design enables multiple and diverse operations. The ability to expand and enhance the modular system is significant. The disadvantages of this system include the necessity for high quality components and the complexity of integrating other systems. An integrated system, on the other hand, represents a 'turnkey' approach which requires the user to execute a simple operation to start the system and obtain products without any user interface. This type of system is useful for selected, repetitive reactions analogous to peptide or oligomeric syntheses. However, because the system is closed, the modification of reactor size, reaction conditions, and integration with peripherals is not possible.

Hardware and Software

Synthesizers. A robust and easy to use reactor is critical to the success of any chemical synthesis. Also, a chemical reactor's material of construction dramatically influences the types of reactions which can be executed in it and, subsequently, the purity profile of the final products (Chemical Resistance Charts, Cole-Parmer Instrument Catalogue, 1995-1996, pp 1672-1680). Typically, peptide synthesis equipment utilizes polyethylene or polypropylene reactors. Although these materials of construction are compatible with the reagent conditions necessary for amide coupling (e.g. N,N-dimethylformamide, diisopropylethylamine, 25°C), they are not compatible with typical cleavage conditions (e.g. hydrofluoric acid or trifluoroacetic acid / dichloromethane). Therefore, most peptide synthesis equipment requires manual transfer and cleavage in a separate glass or teflon reactor. Organic synthesis exploits a broad range of reaction temperatures (e.g. -78 to 200°C) and reagent properties (e.g. acidic, caustic, halogenated, pyrophoric). The selection of reactor construction should be dictated by the chemical and thermal compatibility with the desired reaction conditions. For example, the generation of an imine from a nitrile by a Grignard reaction may require refluxing in THF. These conditions are only compatible with glass or teflon reactors. However, the use of a teflon reactor with some method of heating is hindered by the poor conductance properties of teflon compared to glass. Additionally, by-products and impurities often contaminate the desired products when the reactor material is not compatible with the reaction conditions.

The "DIVERSOMER 8-PIN Synthesizer" (Available commercially from Chemglass, Inc. Vineland, NJ) used for multiple, simultaneous synthesis is represented in Figure 1 (5) . The compatibility of this synthesizer with a wide range of chemical manipulations, including both solution and solid phase chemistry over a broad temperature range, dramatically increases its flexibility. The glass reaction vessels are more compatible with organic synthesis than other materials of construction. Two reactor sizes, for 100 or 800 milligrams of solid support, have also been implemented. The scale of the reaction depends upon the objectives and the nature of the chemistry. This parallel reactor design provides a low-cost opportunity to experience the benefits of more efficient laboratory experimentation including synthetic methodology studies and lead generation, as well as SAR development and lead optimization.

The synthesizer has been described in detail previously (6-8) However, a brief description is necessary to highlight its utility for integration and automation of the system. The unique reactor design employs a fritted vessel, or PIN, to physically contain a solid support (resin). The hollow, fritted glass filter cup serves to contain the solid support, allow efficient mixing between reactants in the reservoir wells and the resin inside the PIN, and facilitate separation of the resin-bound intermediates from excess reagents, solvents, and by-products. The holder block serves to secure the PINS and provides a means for simultaneously manipulating the 8 PINS as a single unit. The reservoir block holds an array of reaction wells which accommodate the fritted glass filters, while concurrently retaining a quantity of reactant necessary to perform the required reactions. Separate reaction wells allow individual reactions to be executed and monitored, while maintaining the integrity of filtrates, intermediates, and final products corresponding to each location in the array. The manifold encloses the upper portion of the PINS, allowing control over the reaction atmosphere. The upper portion of the PINS serve as condensers when a chilled gas is circulated through the manifold, providing a means to maintain reflux. The synthesizer is secured with gaskets and clips to create a gas tight unit. Reagents can be added directly into the PINS by injection through a gasket sealed plate at the top of the manifold.

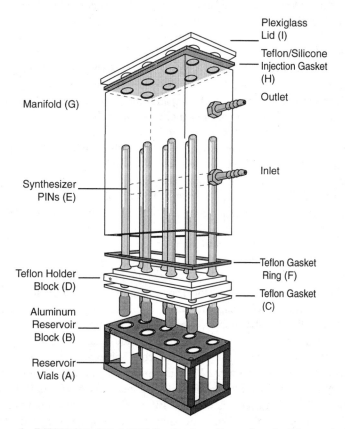

Figure 1. DIVERSOMER 8-PIN Synthesizer (Reproduced with permission from ref. 5. Copyright 1995 Chemglass, Inc.)

Solution phase reactions are executed in an analogous manner using either the fritted PIN or an abbreviated version without the fritted filter at the bottom.

Reaction Equipment.   Although conventional laboratory equipment for agitation and temperature control are sufficient for the one at a time syntheses, equipment compatible with parallel processing is not readily available.   Batch processing and reaction control can be achieved by heating the DIVERSOMER synthesizer on a conventional hotplate or immersion of the reservoir block in a temperature controlled bath in combination with ultrasound or mechanical agitation.

Robotic Automation.   Laboratory automation, although not historically used in synthetic chemistry laboratories, has been implemented in biology, analytical chemistry, clinical chemistry, and process development laboratories.   Two primary methods exist for the introduction and removal of reagents from reactors: a flow-through system employing reactors with inlet and outlet valves or a robotic system with independent reactors.   The integrity of the flow-through system is best achieved by mounting the reactors at a stationary location in the instrument. The number of reactions that can be controlled in this type of system has not exceeded twelve due to the complexity of integrated valving systems.   In contrast, a robotic system enables mobility and modularity of the reactors.

Traditionally, the automation of synthetic chemistry applications has focussed on sequential solution phase synthesis (9-17).   Recently, however, major developments in multiple, simultaneous solid phase peptide synthesis (SPPS) have been reported (18).   Although SPPS reactions are often fully optimized, the generalization of this application of automation is limited since a finite range of synthetic manipulations are employed (e.g. ambient reaction temperatures, standard reagents and repetitive chemistry).

The DIVERSOMER technology fully exploits a wide repertoire of chemical conditions and relies heavily on a flexible automation system.   All liquid sample handling in the DIVERSOMER system is achieved using a customized liquid handling robot (LHR).   The hardware and the software of the LHR have been customized to interface with the DIVERSOMER synthesizer and perform a wide variety of manipulations common to organic synthesis.   Representative modifications to the LHR include 1) the use of custom sample racks, 2) removable plates interfaced with the platform deck, 3) elevation of the deck to enlarge the work envelope in the z direction, and 4) the use of custom probes for injection and slurry transfer.   The robot is currently being used for various tasks involved in the synthetic process including resin loading, reaction cycle monitoring, wash cycles, and parallel purification by Solid Phase Extraction (SPE).

Future refinements are directed toward interfacing the instruments with an industrial robot.   This robot is capable of physically manipulating the

DIVERSOMER synthesizer including placement on the robot deck and movement to reaction and drying stations.

Information Management. As more efficient and productive automated synthesis systems are designed, information management tools for planning, execution, and data handling are necessary. For example, a four step synthesis of 96 compounds simultaneously including four point reaction monitoring at each step and analysis of the final products by two methods could generate more than 1700 discrete pieces of data. Therefore, systems designed to control, communicate, and handle various types of data input and output are necessary.

The DIVERSOMER information system supports both manual and semi-automated synthetic workstations. The Reaction Workbook is the core of this system providing information to the liquid handling robot and other stations in the system. Later systems will include interfaces to pick-and-place robots, reaction equipment and purification and analysis tools. Furthermore, the modular design of the software provides a cost effective way to upgrade to more automated systems.

The complete workstation software includes modules for compound design, reaction information management, control of LHRs and event scheduling. To facilitate the transfer of information between each module, the interfaces between the software programs are well defined. Standardizing the format of the information passed between each module allows the end user to integrate software from different vendors.

Other design goals include the implementation of an easy to use interface to the system for end users which will enable the transfer of information from the scientist to the automated synthetic workstation. The orientation of the software is not directed toward supporting the robotic apparatus associated with the workstation but rather toward enabling the scientist to perform automated parallel synthesis using a familiar paradigm. Rather than input information about the location and volume of a reagent (information necessary for the robotic systems), the chemist enters data about the desired reactions (e.g. reagents selection, quantity, and order and protocol for addition).

Reaction Workbook. The use of a laboratory notebook is integral to a chemist. The documentation of reaction schemes, reagent and solvent parameters, a description of the procedure, and the reaction outcomes are all critical to the execution and reproduction of a synthesis. These lists form the core of the information necessary to drive an automated chemical synthesizer. The multiplication of this information necessary to simultaneously synthesize forty compounds quickly overwhelms traditional methods of notebook information management. For example, a five step synthesis would translate into >200 notebook pages.

The location of each reagent, intermediate, and product must be tracked throughout the entire synthesis process. In other words, where will product "A" be located at the end of the synthetic sequence and where must the reagents be added to create product "A" ? The DIVERSOMER Reaction Workbook, illustrated in Figure 2, automatically assigns well locations to each product and the corresponding reagents.

The key to linking the products with the reagents is a concept which is called the "Product String" in the reaction workbook. A product string is a concatenation of the reagents used to construct the desired compound placed in the order they are used in the reaction sequence. These reagents may only temporarily be a part of the target molecule (e.g. the resin used in a solid phase reaction). For example, the product string Alanine:tButyl:None would identify 3-butyl-5-methyl-2,4-Imidazolidinedione, a hydantoin formed in a three step reaction where the last step, cyclization and cleavage from the resin, did not impart any diversity in the product.

The Reaction Workbook can also accommodate different chemical reactions in a single synthesizer unit by linking reagents and solvents to specific building blocks. For example, both acid chlorides and organic acids could be used to introduce different building blocks by linking the necessary reagents to either the acid chloride or the acid building blocks.

Operation of the System

Reaction Setup. At the beginning of a solution or solid phase synthesis route, the PINS are fitted into the holder block in preparation for the reaction sequence. When using resins, they may be loaded manually or automatically, as a slurry, using the liquid handling robot.

Reaction Cycles. Since organic reactions are often performed under controlled conditions such as a nitrogen atmosphere, reagents or solvents need to be introduced without disruption of the reaction environment. The use of a needle enables automated piercing and injection through a gasket to deliver reagents to the PINS while simultaneously maintaining a controlled atmosphere. This process can also be achieved manually with a syringe before or during the reaction cycle. This is especially attractive while using pyrophoric, unstable, or sensitive reagents that require manual intervention. These types of reagents, common in organic synthesis procedures, are not readily amenable to automation.

Wash Cycles. At the end of each solid phase reaction, the PINS are subjected to a series of wash cycles to remove residual solvents, reagents, and by-products. A typical wash cycle involves several iterations of dispensing fresh solvent directly to PINS using the liquid handling robot. The synthesizer is suspended on the

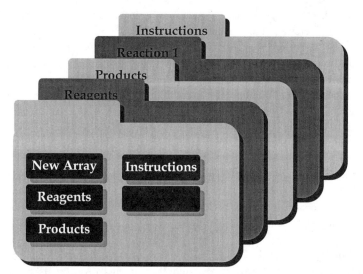

Figure 2. DIVERSOMER Reaction Workbook for Information Management

robot deck, enabling filtrates to be collected or discarded as desired. The user-defined dispense speed for solvent delivery allows for efficient washing of the resin inside the PIN.

Monitoring the efficiency of the wash cycles can be achieved by robotically spotting the filtrates to a thin layer chromatography, or TLC plate and visually confirming the absence of dissolved residues by UV or colorimetric indicators.

Isolation and Purification. Once the final products are cleaved from the solid support, they can be further manipulated in solution. A variety of methods have been developed to affect two-phase extractions. The implementation of SPE methods utilizes cartridges filled with drying agents or chromatography packings enabling drying of the dissolved products or preparative chromatography, respectively. The cartridges are mounted on a manifold at the robot workstation. For reverse phase separations, the manifold is attached to a vacuum box. The LHR dispenses solvents appropriate for column conditioning before transferring the crude products onto each cartridge. Following each transfer, the LHR probe is washed with an appropriate solvent to avoid cross-contamination. Finally, an isocratic or gradient elution protocol is implemented to separate and isolate the desired products. As above, the individual fractions may be analyzed for containment of the product by robotic TLC spotting.

Final Product Handling. Concentration of the final product solutions can be achieved by applying vacuum in a centrifuge device such as a SpeedVac or by maintaining a positive gas flow through a multi-channel manifold. The LHR is essential for the precise transfer of samples between the reservoir block and the SpeedVac rotor.

Following concentration and weighing, the samples are dissolved in solvents amenable to product analysis and biological testing. They can then be quantitatively analyzed by proton NMR and qualitatively by mass spectrometry, GC, HPLC, or TLC. Following final product analyses, the dissolved compounds are transferred to a 96-well plate and these solutions are immediately available for evaluation in biological assays.

Impact of Polystyrene Resins

Exhaustive and quantitative analysis of products from solid phase organic synthesis (SPOS) by proton nuclear magnetic resonance (^{1}H-NMR) with internal or external standards (ISTD or EXTD) in combination with mass spectrum (MS) has revealed impurities resulting from both the polystyrene solid supports and the synthetic route (MacDonald, A. A., Ghosh, S., Hogan, E. M. Kieras, L., DeWitt, S. H., Czarnik, A. W., Ramage, R., *Mol. Diversity*, in press). The reduction or elimination of these contaminants is necessary to avoid "false negatives" within compound libraries.

The selection and use of polystyrene resins is impacted by the presence of entrapped impurities such as solvents, or reaction by-products from copolymerization. These impurities can have a detrimental effect on the reaction sequence if the impurity reacts with, or contaminates, any of the reagents throughout the synthetic scheme. Furthermore, these impurities effectively lower the resin loading and the determination of product yields and reaction monitoring. However, these contaminants are not routinely observed by conventional analysis methods such as thin layer chromatography (TLC), ultra-violet (UV), gas chromatography (GC), or high performance liquid chromatography (HPLC) methods. They can be identified through the use of gel phase ^{13}C NMR or infrared (IR), prior to the execution of a SPOS route. Unfortunately, the impurity profile of the resins is not consistent among commercial suppliers and varies from lot to lot, therefore analysis and/or washing of each resin prior to synthetic use is recommended.

While washing improves the quality of the resins prior to SPOS, by-products resulting from exposure of polystyrene resins to cleavage conditions or reactants from intermediate steps are routinely observed in the final product. All attempts to avoid resin by-products by treating the resins with the final cleavage conditions prior to synthesis were unsuccessful. The nature of these by-products was elusive from the ^{1}H-NMR and MS analyses, and varied dependent upon the functionalized resin and the cleavage conditions. Furthermore, repeated exposures to pre-treatment conditions failed to significantly diminish the isolation of by-products.

Attempts to refunctionalize these resins following pre-treatment proved difficult and the process of pre-treatment resulted in lowering the effective loading capacity of the resin. In many cases, no further chemistry could be successfully implemented on the pre-treated resins. Although resin pre-treatment conclusively identified many of the impurity profiles for a number of our SPOS products, this method of purification is not a recommended method to overcome resin contaminants.

Since adequate procedures to eliminate resin by-products by a resin-based approach have not been identified, product-based purification methods described earlier are recommended. Pre-washing of the resins is advised to eliminate entrapped resin impurities. However, pre-treatment of the resins with cleavage conditions is not recommended as this method did not overcome the poor product profiles and resulted in degradation of the polystyrene resin.

Post cleavage purification methods in combination with quantitative analysis (e.g. ^{1}H-NMR/ISTD or HPLC/EXTD) increases the value of these products and negates the use of compound mixtures or deconvolution strategies. The implementation of parallel purification and quantitative analysis methods with high-throughput synthesis facilitates the integration of chemistry, high-throughput testing, and sample storage and retrieval systems.

Conclusion

The full exploitation of automation in a synthetic chemistry laboratory has not yet been realized. However, the continued application of automation to the routine, repetitive, and error-prone processes for compound synthesis will dramatically impact the productivity and efficiency of the synthetic chemistry laboratory. No single system addresses the needs of every user and the full range of chemical reactions. However, as the demand for such systems increases, the need for new and improved solutions because increasingly evident. It is noteworthy that the execution of combinatorial chemistry strategies can be and, in many cases, is limited by the availability of equipment and instrumentation. For more than one hundred years, synthetic chemists have not been limited by the lack of reaction equipment to generate compounds. Combinatorial chemistry has begun to identify the limitations of existing tools for synthetic chemists. The DIVERSOMER technology provides a set of tools to enable combinatorial and automated synthesis within a modular and open architecture system.

Literature Cited

1.   Gallop, M. A.; Barrett, R. W.; Dower, W. J.; Fodor, S. P. W.; Gordon, E. M., *J. Med. Chem.* **1994**, *37*, 1233-1251.

2.   Gordon, E. M.; Barrett, R. W.; Dower, W. J.; Fodor, S. P. W.; Gallop, M. A., *J. Med. Chem.* **1994**, *37*, 1385-1401.

3.   DeWitt, S. H. In *The Practice of Medicinal Chemistry*; C. G. Wermuth, Ed.; Academic Press Limited: London, U.K., 1996; pp 117-134.

4.   DeWitt, S. H.; Czarnik, A. W., *Curr. Opin. Biotech.* **1995**, *6*, 640-645.

5.   The Chemglass 8-PIN Synthesizer Manual, Version 1.0; Chemglass, Inc.: Vineland, NJ, 1995.

6.   DeWitt, S. H.; Kiely, J. K.; Stankovic, C. J.; Schroeder, M. C.; Cody, D. M. R.; Pavia, M. R., *Proc. Natl. Acad. Sci. USA* **1993**, *90*, 6909-6913.

7.   DeWitt, S. H.; Kiely, J. K.; Pavia, M. R.; Stankovic, C. J.; Schroeder, M. C., *US Patent #5,324,483* **1994**.

8.   DeWitt, S. H.; Schroeder, M. C.; Stankovic, C. J.; Strode, J. E.; Czarnik, A. W., *Drug Dev. Res.* **1994**, *33*, 116-124.

9.   Corkan, L. A.; Plouvier, J. C.; Lindsey, J. S., *Chemom. Intell. Lab. Syst.* **1992**, *17*, 95-105.

10.   Corkan, A.; Lindsey, J. S., *Adv. Lab. Autom. Rob.* **1990**, *6*, 477-97.

11.   Corkan, L. A.; Lindsey, J. S., *Chemom. Intell. Lab. Syst.* **1992**, *17*, 47-74.

12.   Josses, P.; Joux, B.; Barrier, R.; Desmurs, J. R.; Bulliot, H.; Ploquin, Y.; Metivier, P., *Adv. Lab. Autom. Rob.* **1990**, *6*, 463-475.
13.   Lantrip, D. A.; Fuchs, P. L.; Kramer, G. W., *Adv. Lab. Autom. Rob.* **1989**, *5*, 115-137.
14.   Matsuda, R.; Ishibashi, M.; Takeda, Y.,*Chem. Pharm. Bull.* **1988**, *36*, 3512-3518.
15.   Metivier, P.; Josses, P.; Bulliot, H.; Corbet, J. P.; Joux, B., *Chemom. Intell. Lab. Syst.* **1992**, *17*, 137-143.
16.   Sugawara, T.; Kato, S.; Okamoto, S., *J. Auto. Chem.* **1994**, *16*, 33-42.
17.   Weglarz, T. E.; Atkin, S. C., *Adv. Lab. Autom. Rob.* **1990**, *6*, 435-461.
18.   Jung, G.; Beck-Sickinger, A. G., *Angew. Chem. Int. Ed. Engl.* **1992**, *31*, 367-383.

Chapter 20

LiBrain: Software for Automated Design of Exploratory and Targeted Combinatorial Libraries

A. Polinsky, R. D. Feinstein, S. Shi, and A. Kuki

Alanex Corporation, 3550 General Atomics Court, San Diego, CA 92121

LiBrain is a collection of software modules for automated combinatorial library design. Major bottlenecks associated with automating this process have been circumvented through the use of the Chemistry Simulation Engine. This module is trained by chemists to determine the suitability of reactants for a specified reaction, to recognize the risk of undesirable side reactions and to predict the structures of the most likely reaction products. Other modules within LiBrain mine electronic catalogs of chemicals for suitable reactants, incorporate desirable pharmacophoric features into the molecules of the library , and optimize the diversity of designed libraries. LiBrain also facilitates the tracking of library production and characterization, and automatically registers all synthesized compounds. Both maximally diverse exploratory libraries and libraries focused on specific pharmacophores can be designed by LiBrain.

The drug discovery process can be significantly accelerated through the use of combinatorial libraries. Each library typically contains from a few hundred to many thousands of compounds, all synthesized around a common chemical template under similar reaction conditions. As novel biological targets are identified about which no information is available regarding their structural requirements, screening of diverse exploratory libraries can quickly produce valuable "hits". These hits can be further optimized combinatorially through the synthesis of targeted libraries and/or through traditional medicinal chemistry methods. Targeted libraries can also be designed and synthesized when structural information about the target is available from either X-ray or NMR experiments.

The term "combinatorial chemistry" is used to describe a wide variety of disparate methodologies to produce and screen chemical diversity (1-4). Two fundamental features that can be used to differentiate these methodologies are the type of chemistry utilized (solution vs solid phase) and the intended composition of the material to be tested in a bioassay (individual compounds vs mixtures). Relative pros and cons of each approach as applied to small molecule libraries have been recently reviewed (5).

We synthesize small molecule (MW< 600) combinatorial libraries of individual compounds using both solution and solid phase chemistry. There are two major

1054–7487/96/0219$15.00/0

reasons for choosing individual compounds over mixtures. First, the reliability of the screening results decreases drastically as the complexity of the mixtures increases, especially in membrane receptor binding assays. Second, it is very difficult to assess the quality of a library composed of mixtures of compounds to determine whether all the target compounds have actually been synthesized.

The synthesis of libraries of discrete compounds, on the other hand, presents two major challenges. One is to be able to perform very high throughput parallel synthesis to compensate for a lower (compared to mixtures) number of compounds produced. The second is to automate library design and information management so that they can keep up with the high synthesis speed while producing libraries that are optimized with regard to their diversity and/or presence of desired pharmacophoric features.

In this paper, we will present an overview of an integrated library design and chemical information management system - LiBrain - that was developed and is being used at Alanex Corporation to support a high throughput parallel synthesis effort producing approximately ten thousand individual compounds per month.

Bottlenecks in the Library Information Management Process

Several tasks involving chemical information manipulations must be performed in order to design, synthesize and register a combinatorial library. These tasks include the search for suitable reactants from all catalogs of available chemicals, selection of optimal subsets of these suitable reactants so as to produce diverse libraries, planning and tracking library production, and finally, registration of library compounds for subsequent assessment of the structures in the library and SAR analysis. Commercially available tools can be used to partly automate individual steps of this process. The Available Chemicals Directory (ACD) database from MDL Information Systems, and chemical database management systems from MDL, Tripos or Chemical Design Ltd. are examples of such tools. However, when using these tools, a chemist must be constantly involved in the handling of frequently occurring special situations and exceptions. For example, after obtaining a list of reactants with desirable reactive functional groups through an ACD database substructure search, the chemist usually manually removes reactants where the target functional group is found in a wrong context (e.g. a search for ketones will yield α and β diketones), or reactants that might cause side reactions, or reactants that are simply too expensive. Another example would be choosing the most likely reaction product when groups with different reactivities are present in one of the reactants or when the regiospecificity of the reaction is influenced by the relative steric bulk of the reactants.

This need for constant exception handling in various stages of library design and production is a major bottleneck which makes the automation of the whole library information management process a challenging task. LiBrain resolves this bottleneck through the precise use of the concept of *virtual libraries* whose content is created and controlled by a *Chemistry Simulation Engine* contained within LiBrain.

LiBrain Foundations

Virtual Libraries. The term "virtual library" is widely used to describe an electronic representation of a library of computer generated chemical structures. Two of the first examples of such libraries were TRIAD, a collection of tricyclic hydrocarbon structures, and ILIAD, a collection of acyclic molecules, used by CAVEAT for *de novo* molecular design (*6*). In the context of combinatorial chemistry, virtual libraries typically represent all products that can be obtained through a specified chemical reaction using a pre-selected set of reactants. Products contained in the virtual library can be evaluated using an appropriate scoring function prior to the synthesis, and the best subset can then be selected and produced.

LiBrain's virtual libraries refer to electronic representations of libraries that consist *only* of molecules synthesizable by chemical reactions optimized for the Alanex high throughput parallel synthesis equipment. This makes the tight integration of library design and library production possible, because any and all structures selected from the virtual library can be quickly synthesized using available reagents. Reactants are considered available if they are listed either in one of the databases of commercially available chemicals, or in the corporate database of proprietary precursors. Taking into consideration all currently available reactants makes LiBrain's virtual libraries huge (tens of millions of products for some reactions) which is beneficial for the design of both very diverse and very targeted libraries. On the other hand, not only would it be impractical to enumerate these virtual libraries explicitly and store them on disk due to their size, but it would also not be desirable. The precise content of the virtual libraries is constantly changing as suppliers add or delete chemicals in their catalogs, as new proprietary precursors are synthesized in-house, or as some of these precursors are used up. To work automatically with such dynamic virtual libraries, LiBrain requires tools that can access and process the most up-to-date versions of reactant databases, select (as chemists would do) sets of suitable reactants with minimized risk of side reactions, and generate (on the fly) and evaluate subsets of virtual products, handling special cases as chemists would. These tools within LiBrain are based on a core module, the Chemistry Simulation Engine.

Chemistry Simulation Engine. The Chemistry Simulation Engine is a software module that performs three major tasks. First, given the structure of a reactant, it evaluates if that reactant is capable of participating in a specified chemical reaction. Second, given multiple reactant structures, it can determine whether a specified chemical reaction can occur among them and whether there is a potential for undesirable side reactions. Third, for multiple reactants capable of participating in a specified reaction, it can generate the structure of one (the most probable, or "intended") product, or alternatively all possible products. The Chemistry Simulation Engine is written in Procedural Language (PL), a modified Pascal incorporated into MDL's ISIS suite of chemical information management programs (7).

The Chemistry Simulation Engine is built around an extensible Chemistry Knowledge Base that contains the detailed description of each chemical reaction optimized for the high throughput parallel synthesis. This knowledge base is implemented as a specially configured ISIS database. A precise description of each chemical transformation is given by the reaction scheme, substructures of the reacting groups and a mapping scheme indicating which reactant atoms become part of the product. If the library synthesis involves several steps, each step is described separately, then the steps are chained together.

Information for the automated generation of the database search queries for reactants is supplemented by the list of "errant" substructures that are likely to be confused with the target substructure by the search engine. For example, if one of the reacting groups is a primary amine, the substructures of a primary amide and thioamide, guanidino group, amidino group, monosubstituted urea, etc., would be included into the "errants" list.

To evaluate the risk of side reactions, a list of functional groups that do not participate in the desired reaction, but which can react with one of the participating groups under the specific reaction conditions is recorded. For example, in the reductive amination reaction, this list of groups that can potentially react with the amino component includes isocyanate, acyl- and sulfonyl chlorides, epoxy group, etc.

Differences in reactivity for similar functional groups are described through the creation of variations of the general reaction and the prioritization of these variations according to the relative reactivity of the participating groups. For example, for the simple reaction of a carboxylic acid with an amine, four variations are created. These depict a carboxylic acid reacting with, in order of decreasing priority, a primary alkyl amine, a secondary alkyl amine, a primary aryl amine and a secondary aryl amine. The facile creation of reaction variations provides chemists with a flexible tool to describe any special case situation that could be relevant for a given reaction.

All information contained within the Chemistry Knowledge Base is provided by the chemists who optimized the given reactions for the parallel synthesis. In a sense, chemists *teach* Chemistry Simulation Engine the details of each specific chemistry. Aware of these details, Chemistry Simulation Engine becomes capable of handling special cases and exceptions as mentioned above. The Chemistry Knowledge Base is constantly updated as experience in running specific reactions is accumulated.

Molecular Representation and Diversity. The viable reactants are readily separated by the Chemistry Simulation Engine, using the knowledge base mentioned above, into the variable R-group portion(s) and the obligatory reacting functionality. The variable functional groups are in the R-group portion, and these are the variable features which will be presented on the surface of the resultant combinatorial products. Therefore, a diversity in R-group features will imply a diversity in product features.

Diversity optimization for both exploratory and focused library design requires dissimilarity calculations. The result of these calculations depends heavily on the choice of features used to compare structures with each other, and on the way in which these features are encoded into a feature vector describing properties of the R-groups. We assume that combinatorial libraries must be diverse with respect to structural elements that are typically found to be responsible for binding affinity to biomolecular targets. We call these elements the affinity elements. Affinity elements should be treated as closely related or completely different according to how strongly their effects on affinity are correlated. Those affinity elements that tend to function interchangeably with respect to binding affinity belong to the same class of affinity elements. LiBrain considers seven such classes: aliphatic hydrophobic, aromatic hydrophobic, basic, acidic, hydrogen bond donor, hydrogen bond acceptor, and polarizable heteroatom.

Each R-group may have multiple affinity elements belonging to different classes. Then the simplest molecular representation based on affinity elements would be through a feature vector that has as many components as there are classes, and the value of each component would be 1 or 0 depending on the presence or absence of the affinity element of a corresponding class in the R-group. This representation would be analogous to a binary key approach employed in database searching, only with many fewer bits.

The molecular representation described above is highly oversimplified and leads to the loss of important distinctive features of R-groups. LiBrain uses an extension of this molecular representation that preserves the focus on large differences in modes of interaction with the target (e.g. hydrophobic vs. ion pairing). At the same time, an additional three levels of detail are added to the description of R-groups through additional feature vector components accounting for the total number of affinity elements within the same class, their topological location and selected affinity related physical properties. In LiBrain's molecular representation, each class is described by several components, typically four or five, which leads to a total of 28 (7x4) or 35 (7x5) components in a feature vector.

A central purpose of library synthesis and screening is to obtain new patterns of affinity elements as presented by a substantially new lead structure. The reactant feature vectors built up from several components for each affinity element class are designed to have the level of discrimination appropriate to describe the significantly different patterns relevant to exploratory lead discovery. The LiBrain diversity optimization benefits from this focus on the use of a dissimilarity measurement that is oriented towards exploratory discovery.

LiBrain Components and Organization.

Tasks. LiBrain has been designed to accelerate several phases of the drug discovery process. First, at the early stage of the project when nothing is known about the structural requirements of the target, initial lead identification is aided by LiBrain's ability to design maximally diverse exploratory libraries. Second, generation of structurally diverse compounds related to an identified lead is aided by LiBrain's ability to apply structural constraints to library design based on fragmentary knowledge of the pharmacophore(s) involved. Third, performing combinatorial optimization or fast analoging is facilitated by LiBrain's ability to select diverse sets of substituents.

Designing libraries means selecting small subsets of the large virtual library for synthesis. LiBrain is built around a sequence of modules each performing an analysis of reactants or reaction products with successively increasing level of discrimination. The type of the designed library (e.g. exploratory or targeted) determines the order in which the modules are applied and the criteria used for reactant selection.

The modules that make up LiBrain are shown in Figure 1. The successive decrease of the size of the virtual product matrix after application of LiBrain's modules is shown in Figure 2. The Chemistry Simulation Engine at the foundation of LiBrain ensures that all combinatorial products considered are rapidly accessible using one of the reactions optimized for parallel synthesis and available chemicals. The Automatic Reactant Mining (ARM) module identifies and retrieves all reactants suitable for a specified reaction (P0 → P1). Each such compatible reactant is then considered as an R-group which is the molecular structure of the reactant with its reacting functionality removed. Chemical features of the R-group are then analyzed and recorded by the Affinity Element Analysis module. Reactants that do not possess the desirable features can be filtered out at this point by the Reactant Composition Filter module (P1 → P2). The Diversity Optimization Module selects maximally diverse subsets of reactants from the filtered reactant lists (P2 → P3). After a considerably reduced number of reactants are thereby selected, explicit product formation and evaluation can be performed by the Virtual Product Evaluation module, and the final optimized subset is chosen (P3 → P4). When an actual library has been produced, all synthesized compounds are automatically registered using the Product Registration module.

Specific sizes for the product matrices at the various stages in the process are discussed below within the description of each module. Together, these modules enable a wide variety of library designs to be carried out.

Automatic Reactant Mining (ARM). From the point of view of combinatorial chemistry, all available compounds may be viewed as a rich reservoir holding potentially valuable chemical feed stocks for library synthesis. Computational methods need to begin with the same viewpoint, and indeed should, if well designed, be able to rapidly amplify the chemist's ability to retrieve comprehensive sets of viable reactants.

The number of commercially available compounds is large. For example, the ACD database contains in excess of 10^5 molecules. In-house synthesis provides additional compounds for use as special combinatorial starting materials. The straightforward application of queries for molecules containing a desired reactive functional group to such a large body of potential reactants often results in the retrieval of too many compounds. On closer examination, many turn out to be unsuitable for high yield parallel chemistry. The accurate recognition of numerous problematic reactants is achieved by employing detailed reaction specific information in the Chemistry Knowledge Base on differential reactivity and on predictable side reaction and polymerization mechanisms.

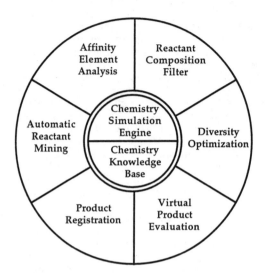

Figure 1. LiBrain components.

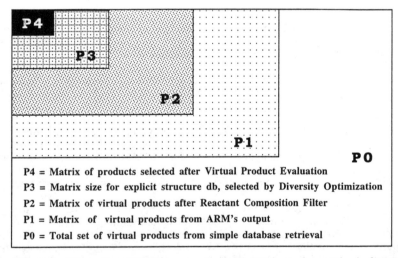

P4 = Matrix of products selected after Virtual Product Evaluation
P3 = Matrix size for explicit structure db, selected by Diversity Optimization
P2 = Matrix of virtual products after Reactant Composition Filter
P1 = Matrix of virtual products from ARM's output
P0 = Total set of virtual products from simple database retrieval

Figure 2. Illustration of the process of selecting subsets for synthesis from the original vast virtual library.

The ARM module is designed to perform automatic database searches and produces lists of suitable reactants with a minimized risk of side reactions. ARM begins by screening the database using a variety of user specified criteria. These criteria include concerns regarding preferred vendors, price, purity, acceptable salts and undesirable elements and isotopes. ARM then retrieves the information regarding a user specified chemistry from the Chemistry Knowledge Base, then retrieves all molecules with the reactive functional groups required for the reactant(s). Compounds are then eliminated if they contain excessive, undesirable, or unstable functional groups. Molecules that contain both moieties involved in the reaction are also removed as polymerization threats.

Once the reactants have been characterized, user defined rules are applied to aid in the pruning of the reactant lists. A rule that is commonly employed is to limit hits to those compounds with only a single "most reactive" group. For example, assuming that the chemistry involved differentiates aryl amines from alkyl amines, then compounds with one of each would be retained, while a compound with two alkyl amines (which have the greatest reactivity) would be discarded. Likewise, a compound with two aryl amines only would also be discarded. This has the effect of preventing the formation of a mixture of products.

Table I illustrates how the ARM module prunes an exhaustive list of potential reactants. The retrieval of reactants for reductive amination and urea formation from isocyanates is shown. Both reactions can incorporate primary/secondary aliphatic/aromatic amines. When one simply retrieves these amines from, for example, the Available Chemicals Directory (ACD) database, the list of such amines includes over 17,000 molecules. Retrieving aldehydes and ketones by direct substructure search will similarly yield over 20,000 molecules. However, most of these molecules are unsatisfactory because they fail to meet the criteria mentioned above. For the reductive amination and urea formation reactions, 70% and 72% of the amines are rejected, respectively. More are rejected for the urea formation reaction because secondary arylamines are not considered suitable reactants.

Table I. Size of virtual product matrices before and after using ARM

Reaction	Components	P0	P1
Reductive Amination	Ald./Ket. x Amines	20,663 x 17,595 (360 million)	5733 x 5187 (30 million)
Urea Formation from Isocyanate	Isocyanate x Amines	260 x 17,595 (4.6 million)	177 x 4913 (870 thousand)

The virtual library size using only the suitable reactants for these two reactions (P1) is still huge. The direct enumeration of these virtual products in an explicit database is not feasible, particularly as more than 30 different combinatorial reactions are currently in use with more being developed continually. Further reductions in size (P1->P2->P3) are performed by the Affinity Element Analysis, Reactant Composition Filter and Diversity Optimization modules, described below.

Affinity Element Analysis. The suitable reactants retrieved by ARM, according to the precise and detailed criteria in the Chemistry Knowledge Base, are sent in the form of a pair (or more) of reactant lists to the LiBrain Affinity Element Analysis module. These lists are transformed by an automatic analysis of the reactant molecular structures into sets of reactant feature vectors which will provide the basis for reactant selection and dissimilarity calculations.

The first step is to determine which affinity elements are present and their topological location. The seven affinity element classes differentiate the R-group substructures as aliphatic hydrophobic, aromatic hydrophobic, basic, acidic, hydrogen-bond donor, hydrogen-bond acceptor, and polarizable heteroatom. For the two hydrophobic classes, LiBrain tracks the topological distances of each non-hydrogen atom, recorded as the number of bonds away from the connection atom for that R-group, whereas for the other classes the topological distance of each functional group found is recorded. The number of bonds provides valuable information on how far outward from the product core the affinity element is extended, which reflects the varying shape of R-groups. For each R-group, additional discriminating information is also preserved within each affinity element class. For example, for each affinity element of the class "basic", a chemical property (pKa) is specified.

The Affinity Element Analysis module thus translates each R-group into a feature vector which for each class specifies the total number of affinity elements in this class, topological distance information for the affinity elements, and chemical properties which reflect, and enable discrimination between, different affinity elements belonging to the same class. If we employ four vector components for each affinity element class, this yields a 28-dimensional reactant feature vector, $\vec{R}$. In the comparison of two molecules, each of the seven classes will then make four independent contributions, referred to as the Dissimilarity Terms, to the overall dissimilarity measure.

$$\text{Dissimilarity Distance} = \sqrt{\sum_{\alpha=1}^{D} \left(\text{Dissimilarity Term}_\alpha\right)^2} = \left| \vec{R}_i - \vec{R}_j \right|$$

The components in $\vec{R}$ synopsize in a systematic way the affinity elements present by first specifying in one vector component the total number of hydrophobic atoms or polar functional groups present in that affinity element class (multiplied by a scale normalization factor). The next vector component specifies additional physical property information on the affinity elements within the same class, then the remaining vector components characterize the topological distribution of these atoms/functional groups within that class. As a specific example, in the case of acids and, separately, bases, the average pKa can yield the second vector component, and the last two components of the reactant feature vector from this class may be assigned as the average and maximum bond count in the topological distance distribution of the possibly several acids (or bases) present.

Reactant Composition Filter. The reactant feature vectors are also directly used to assess for the presence or absence of desirable chemical functional group classes in targeted phases of LiBrain library design, in which selected functional groups are enriched in the planned library. The Reactant Composition Filter very quickly applies criteria that require that at least one of the reactants bears a desired affinity element, or that neither reactant possesses a certain affinity element, or that neither reactant has more than a specified number of a certain affinity elements, or that the R-group sizes be in a certain range, etc. In addition, the Reactant Composition Filter has the enhanced capability to accept or reject according to the topological information on the affinity elements as expressed in the reactant feature vectors. The output is the correspondingly pruned subsets of the P1 reactant lists; these lists of filtered reactants then defines P2. The LiBrain reactant analysis can be directed towards more or less targeted diversity as desired - or as justified by the level of information garnered in the specific project - simply by adjusting the affinity element composition criterion to more or less restrictive cutoffs.

Diversity Optimization. The basis of the LiBrain diversity optimization is i) the dissimilarity "distance" between compounds calculated from the reactant feature vectors generated by the Affinity Element Analysis, and ii) the maximin method of generating a Maximum Diversity Trajectory. The dissimilarity distance and the concept of the Maximum Diversity Trajectory is described here; the details of the maximin algorithm which yields this trajectory are given in the Appendix. Broadly speaking, the method used in LiBrain applies a dissimilarity based diversity optimization approach.

What do we mean by optimizing "diversity"? Diversity is often used to describe i) a large separation or dissimilarity between structures within a library, or ii) an unbiased and random distribution, or iii) to indicate that a high degree of coverage of the R-group space has been achieved to some extent with a library. What do we mean by optimizing "diversity"? Here we will focus on obtaining a <u>well-dispersed</u> distribution, which corresponds to a certain <u>uniformity</u> of coverage in an R-group space, distinct from random sampling. As discussed below, this is equivalent to minimizing, for any given number of reactants, the size of the largest *hole* present in the distribution. The diversity maximization algorithm selected, a Maximum Diversity Trajectory algorithm, is a variant of the general approach known as the "maximum dissimilarity approach" (*8*), and is specified in the Appendix. These approaches rely only on balancing the *relative* distances between compounds in the representational space, and operate equally well with relatively sparse to relatively crowded sets.

The balancing of relative distances between compounds is distinct from random sampling, since in the latter case the subsequent samples are completely oblivious to the previous samples, whereas in the former (desirable) case the subsequent samples are designed to fill the largest holes and thus create a well-dispersed set. This contrast is illustrated in Figure 3, a 2-dimensional example of subset selection by the Maximum Diversity Trajectory algorithm.

The diversity algorithms within LiBrain currently concentrate on optimized reactant selection for a single library based on a single reaction, and hence containing a common product core or scaffold. Clearly, enhancing diversity through the use of many core structures or templates is an essential further strategy in drug discovery.

The Maximum Diversity Trajectory Approach. Given the very large size of typical virtual product matrices, it is critical to first reduce the proportions of the matrix by selecting subsets of the reactants. It is natural to select these subsets in such a way so as to optimize their diversity. Reactant diversity optimization is hence a filter and an enhancement which can precede either synthesis or further refinement through more detailed virtual product evaluation.

The key aspect is that the diversity optimization is implemented not by selecting a subset of predefined size, such as 1.0% or 10% of N_{total}, but rather is implemented through a certain sequencing or prioritization of the total original reactant list, as detailed in the Appendix. We say that this optimization defines a "Maximum Diversity Trajectory", which means that for any value of $N < N_{total}$, the first N reactants will produce a maximally well-dispersed subset of size N. As N is increased, the subsequent samples progressively fill in the smaller and smaller holes remaining to yield successively finer-grain uniformity in coverage of the R-group space. The advantage is that there is no need in advance to specify the level or fineness of the desired sampling. The Maximum Diversity Trajectory thus calculated provides a practical and beneficial *continuous scale progression* from the early coarse and well-dispersed distribution to a later fine and well-dispersed distribution.

Virtual Product Evaluation. Referring to Figure 2, the output of the Affinity Element Analysis and Diversity Optimization modules at a specified matrix size is a pair (or more) of reactant lists which define the subset P3 in the virtual product space. Generation of the explicit virtual product structures for P3 is performed by the Virtual Product Evaluation module. This module uses the detailed reaction rules contained within the Chemistry Knowledge Base and produces databases of virtual products. These databases are amenable to processing and evaluation using any of the tools available through the MDL ISIS environment. Typically, 3D pharmacophore searches are performed, both rigid and flexible.

Automatic Product Registration. Due to the high throughput of the combinatorial chemistry process, automation at this stage is vital to information management. The registration process begins with an "electronic notebook" wherein chemists provide a description of an actual synthetic array they have produced. The Chemistry Simulation Engine then retrieves the structures of the reactants used and generates the structures of the intended products and a list of masses corresponding to possible side products for each well in the library. The Product Registration module combines these structures and mass lists with library tracking information and analytical data (e.g. mass spectra), and then imports this library into the corporate combinatorial chemistry product database. This product database is also maintained and manipulated within the ISIS environment, allowing for its integration into the LiBrain library design process.

The generation of molecular weight data for the intended and side products allows for further automation in the maintenance of the product database by the Product Registration module. In the event that mass spectral data suggests the presence of products other than those intended, an automated comparison to the list of possible side product masses can suggest the identity of the molecules in question. Furthermore, the incorporation of previously synthesized products into the library design process can be limited to those for which mass spectral data confirm the presence of the intended products.

Library Design Using LiBrain

Two of the most important applications of LiBrain are i) the design of maximally diverse exploratory libraries and ii) the design of targeted libraries using available information, even if it is preliminary and fragmentary, about the required pharmacophoric features. In both cases, the ARM is first applied to select reactants suitable for a specified chemistry, and the Affinity Element Analysis is performed to characterize the reactant's features.

Exploratory libraries. In the design of a library with the maximum possible diversity, the Reactant Composition Filter module is either not applied at all, or applied with very general selection criteria. For example, the molecular weight limit can be set to reject very large reactants. The Diversity Optimization module is then used to select the most diverse subsets of suitable reactants. The size of the virtual product matrix at this step is matched, by the use of Maximum Diversity Trajectory, to the planned size of the corresponding library (P3 = P4), and all compounds from this matrix are synthesized. Typically, the size of synthesized libraries is between 10^3 and 10^4.

Targeted libraries. When it is desirable to use available information about the required pharmacophoric features, the Reactant Composition Filter module is used to select reactants which contain the desired features. The Reactant Composition Filter also rejects reactants with affinity elements that must not be present in the product, or reactants possessing more than the maximum number allowed for a given affinity element. In this case, the Diversity Optimization module is used to reduce the virtual product matrix to the size that makes exhaustive virtual product formation and evaluation computationally feasible. This size (P3) is typically in the range between 10^4 and 10^5. As the last step, the Virtual Product Evaluation module is applied and the size of the virtual product matrix (P4) is brought down to 10^3 -10^4 for subsequent synthesis.

Conclusions

High speed combinatorial parallel synthesis at a rate of tens of thousands of compounds per month demands an automated library design and information management system. The LiBrain software addresses this challenge and provides combinatorial chemists with efficient tools for designing and tracking combinatorial libraries. Such automation has been made possible through the use of the Chemistry Simulation Engine, a module that is trained by chemists to find reactants with minimized risk of side reactions and then select optimal sets of reactants for the synthesis of either exploratory libraries with maximized diversity, or targeted libraries that are enriched with compounds with desirable pharmacophoric features.

Appendix: Maximum Diversity Trajectory Algorithm

We define the precise variant of the dissimilarity-based diversity method employed within the Diversity Optimization module, and present results from illustrative runs in 5 dimensions. This includes the introduction of a useful direct measure of the length scale, Λ , at which a given library is being refined as new compounds are selected. This algorithm in LiBrain is typically used in reactant representational spaces with dimensions on the order of 20-40.

Maximum Diversity Trajectory Algorithm -- Maximin. Suppose we have N_{total} points in a D-dimensional working space W of unit volume, spanning from 0 to 1 in each dimension. It is further assumed that we start with a set, S(m), of m points, m>0, which has already been preselected either based on an existing library production schedule or simply with first point (m=1) selected intelligently. Our task is to choose additional (N_{total} - m) points such that the resultant sequence of sets S(N), with $m<N<N_{total}$, has a maximum diversity, i.e., the points in the set S(N) are always optimally well-dispersed over the whole space W.

The dissimilarity between two reactants, i and j, is measured by a distance function, d_{ij} , in the representational hyperspace with

$$d_{ij} = \left| \vec{R}_i - \vec{R}_j \right| \tag{A1}$$

where $\vec{R}_i$ is a D-dimensional vector representing reactant i. Naturally, the dissimilarity between an additional reactant j and the set of N reactants already selected, S(N), may be described by the minimum distance $d_{ext}^{min}\left[j, S(N)\right]$ of point j to the set of N points in the hyperspace:

$$d_{ext}^{min}\left[j, S(N)\right] = Min\left[d_{ij}\right] \text{ for all points } i \in S(N) \text{ , ext signifies: } j \notin S(N) \tag{A2}$$

All points j which are external to the set S(N) are candidates to become the (N+1)th member of the Maximum Diversity Trajectory. In order to achieve a maximum diversity, therefore, each newly chosen point k should be selected from among the candidates j to have the largest minimum distance, $d_{ext}^{min}\left[k, S(N)\right]$, to the set S(N)

500 / 5000 Molecule Points in 2D

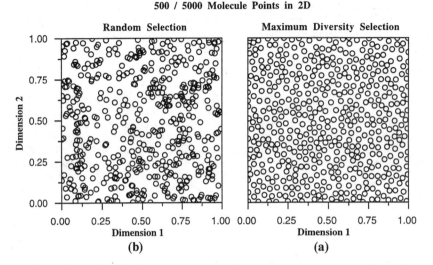

Figure 3. Comparison of two sets of 500 sampling points selected from a randomly generated pool of 5000 points in a 2-dimensional space. The set (a) is given with the present *maximin* algorithm and set (b) is selected randomly. It is seen that the *maximin* algorithm yields sampling points which are more uniformly distributed.

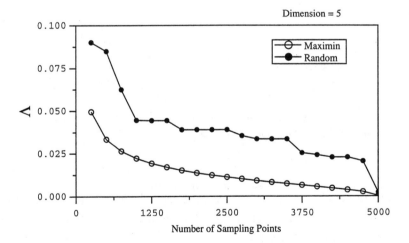

Figure 4. Comparison of Λ, the size of "largest remaining hole", in the sets S(N) selected with two different sampling algorithms: (a) *Maximin*; (b) Random, from a randomly generated pool of 5000 points in a 5-dimensional unit volume hyperspace. The coverage of a given library can be assessed by monitoring this length scale Λ, and the maximum reduction in Λ at the minimum size N is a desirable property conferred by the Maximum Diversity Trajectory employed here.

which has already been chosen. That is, to extend the set S(N) to S(N+1) one should always choose the next additional point k such that

$$d_{ext}^{min}\left[k, S(N)\right] = Max\left[d_{ext}^{min}\left[j, S(N)\right]\right] \text{ for all points } j \notin S(N) \qquad (A3)$$

To see the idea of Eq. A3 another way, a small value of $d_{ext}^{min}\left[j, S(N)\right]$ means that that j is a close neighbor of at least one of the existing points in S(N), and close neighbors are to be avoided.

The value of the $d_{ext}^{min}\left[k, S(N)\right]$ for the next point to be selected by the criterion Eq. A3 approximates the size of the largest hyperspherical hole in the set S(N). This is because for a given set S(N) of N points in the bounded hyperspace, the center, c_{max} , of the largest hyperspherical hole of the set S(N) has a maximum value of

$d_{ext}^{min}\left[\vec{R} \in W, S(N)\right]$ compared to any point in the space W.   At all times, the centers

of holes in the set S(N) are local maxima in $d_{ext}^{min}\left[\vec{R} \in W, S(N)\right]$ with a value of

$d_{ext}^{min}\left[\vec{R} \in W, S(N)\right]$ equal to the radius of that hole. So long as there are remaining points available near c_{max} , the *maximin* algorithm will select as the next additional point that which is closest to the center of the largest hole in the set S(N).   For *any* series of sets S(N), generated by any algorithm, we define this approximate largest hyperspherical hole size as the primary length scale, Λ , which characterizes the progress of the sampling:

$$\Lambda(N) \equiv Max\left[d_{ext}^{min}\left[j, S(N)\right]\right] \text{ for all points } j \notin S(N) \text{ for any series of } S(N) \qquad (A4)$$

In Figure 4, $\Lambda(N)$ is plotted as functions of N for two sequences of selected sets S(N) with two different selection algorithms: (a) Maximin (b) Random. In contrast with the random selection, which gives an erratic decrease of $\Lambda(N)$, the *maximin* algorithm generates sampling sets S(N) with smoothly   decreasing $\Lambda(N)$. The stepwise decrease of $\Lambda(N)$ indicates that many   points selected by the random algorithm do not improve the diversity of the set S(N). As a result of the systematic length scale improvement generated by the *maximin* algorithm, the corresponding $\Lambda(N)$ are smaller than those for random selection for all N< N_{total}, and especially for small N.

This *maximin* algorithm can be implemented very efficiently since for the (N+1)th point there is a recurrence relation

$$d_{ext}^{min}\left[j, S(N)\right] = Min\left[d_{j,k(N)}, d_{ext}^{min}\left[j, S(N-1)\right]\right] \qquad (A6)$$

where k(N) denotes the immediately previous chosen point, which is the Nth point in the set S(N).   As a result, choosing N points out of a total N_{total} points requires the

calculation of only $N\left[N_{total} - (N-1)/2\right]$ distances and twice as many comparison operations. For example, it takes ~3 minutes to select an optimally diverse sequence of 10000 points from 10000 total points in a 5-dimensional hyperspace on a SGI (200MHz R4400).

In addition to its simplicity and efficiency, the present approach, unlike methods based on cluster analysis, does not require prescribed tolerance parameters for clustering. It can, in one run, provide a reactant selection of continuously varying size N with maximal overall diversity at any given stage during the sequence to meet different design objectives. The present algorithm in a simple way guarantees that each additional point will be most dissimilar to the set of points that has already been chosen, and through the resultant evolution yields the desired Maximum Diversity Trajectory whose remaining holes (as measured by Λ) smoothly and continuously shrink in size. Therefore we refer to the resulting sequence by the term Maximum Diversity Trajectory. The analogous point *generation* algorithm in Monte Carlo methods which yields a very similar non-random and well-dispersed distribution is the "quasi-random" numerical algorithm (*9*).

Philosophically, the present maximin algorithm and those based on "D-optimal design" (*10*) which has been used in recent work(*11*), have the same spirit that one should choose the new point where the least is known about the property (e.g. receptor binding) concerned. However the D-optimal algorithm is complex and not very efficient since it involves matrix inversion and iterative optimization, and like other iterative procedures will have difficulty finding the global optimum when the original pool is huge.

References

1. Madden, D.; Krchnák, V.; Lebl, M. *Perspectives in Drug Discovery and Design* **1994**, *2* , 269-285.
2. Gallop, M. A.; Barrett, R.W.; Dower, W.J.; Fodor, S.P.A.; Gordon, E.M. *J.Med.Chem.* **1994**, *37* , 1233-1251.
3. Gordon, E.M.; Barrett, R.W.; Dower, W.J.; Fodor, S.P.A.; Gallop, M. A. *J.Med.Chem.* **1994**, *37* , 1385-1401.
4. Furka, A. *Drug Development Research* **1995**, *36* , 1-12.
5. Terrett, N.K.; Gardner, M.; Gordon, D.W.; Kobylecki, R.J.; Steele, J. *Tetrahedron* **1995**, *51*, 8135-8173.
6. Lauri, G.; Bartlett, P.A. *J. Comp.-Aided Molec. Design* **1994**, *8* , 51-66.
7. ISIS/Base 1.2, ISIS/Host 1.3.1; MDL Information Systems, 1995.
8. Bawden, D. In: *Chemical Structures 2. The Intenational Language of Chemistry* ; Warr, W. A., Ed.; Springer-Verlag, Heidelberg, 1993; pp 383-388.
9. Press, W.H.; Teukolsky, S. A. *Computers in Physics* **1989**, *3* , 76-79.
10. Fedorov, V. V. *Theory of optimal experiments* ; Academic Press, New York, NY, 1972.
11. Martin, E.J.; Blaney, J.M.; Siani M.A.; Spellmeyer, D. C.; Wong, A. K.; Moos, W. H. *J.Med.Chem.* **1995**, *38*, 1431-1436.

Chapter 21

Automation Issues at the Interface Between Combinatorial Chemistry and High-Throughput Screening

Barr E. Bauer

MDL Information Systems, Inc., 14600 Catalina Street,
San Leandro, CA 94577

Abstract: Combinatorial Chemistry and High Throughput Screening are incompletely integrated to take full advantage of the emerging small molecule synthesizer technologies and achieve the promise of high sample and data volumes. Integration will require new technologies (automated quality control instrumentation), new decision support software with expanded control scope and intermodule feedback, and changes in established registration procedures. In addition, performance bottlenecks introduced by this automation must be managed. This paper will explore the automation issues required for successful high throughput discovery.

The advent of high throughput screening (HTS) and combinatorial chemistry as enabling bioagent discovery technologies has raised the expectation that large numbers of new chemical entities will be created and biologically evaluated in a continuous, factory-like process. The emerging small molecule automated synthesizer technologies raises the expectations even further by providing the means by which vast numbers of new chemical entities can be generated.

In fact, a technological and performance gap exists between combinatorial chemistry and HTS technologies. The technological gap is only partially bridged by the automated synthesizer technology. The ability to create new chemical entities under programmatic control brings with it new responsibilities for automated quality control and information feedback to ensure that the entire process delivers the expected new entities. Likewise, the performance gap has several process bottlenecks that must be overcome to ensure a sufficiently high rate of sample flow through the process.

1054–7487/96/0233$15.00/0

The challenge that faces many research organizations is the rapid and successful implementation of these multiple new technologies and to successfully bridge the gap. The result is an integrated bioagent discovery process that resembles a manufacturing process more than a traditional bioagent discovery process. Key to success will be the introduction of automation and intracomponent feedback capabilities. Additionally, the removal of the obvious and nonobvious process bottlenecks, selective researcher intervention in the process, and changes in some time honored processes are also essential to keeping a high rate of sample creation while maintaining quality.

This paper will define possible future requirements for both a high throughput discovery process and their component technologies. These technologies either do not exist today or are in their infancy. The process, however, is definable against clearly stated objectives of many companies actively moving into the arena of combinatorial chemistry and HTS. The discussion will provide a process-based solution that will frame the requirements for high throughput discovery and present technological challenges.

Paradigm Shift

Before discussing the gap and its issues, it is useful to look at the big picture to position the various processes and gap. In short, the classical one-loop optimize-screen-analyze paradigm for bioagent discovery has been transformed into a new paradigm that introduces a new loop containing HTS and combinatorial chemistry (Fig. 1). The main purpose of this second loop is to overcome the problem of starting points for optimization. This problem worsens as molecular biology techniques have increased the rate of new screening targets and can generate targets for which small molecule intervention is unknown. The problem is magnified when competitive and cost pressures are considered.

Interrelationship of Enabling Technologies

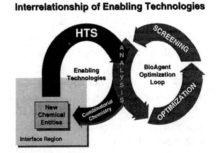

Figure 1. How combinatorial chemistry and HTS interrelate
to the established process for bioagent discovery.

The cornerstone of the new loop is novel screenable chemical entities. These entities today are generally small molecules as opposed to peptides or nucleotides. Characterization is a requirement for either discrete molecules or mixtures. The samples are frequently prepared on microwell plates for both storage and screening. Finally, the chemical entities are often associated into a library, designed and managed with combinatorial chemistry process management and analysis software intended to explore defined regions of SAR.

The basic assumption is that the second loop must rapidly generate and screen samples resulting in a steady stream of hits advancing into the classical loop or hits advancing back into the combinatorial chemistry software acting as seeds for further rounds of optimization. Any bottlenecks in the second loop, specifically around the region where new entities are generated ultimately impedes overall system performance and slows the rate of data generation and hit advancement. This potential bottleneck region, referred to earlier as a gap and highlighted in Figure 1 is where a new process and new technology will be required.

Completing the High Throughput Discovery Process

The gap-spanning process that will be the focus of the remainder of this article is shown in Figure 2. It's mission is to transform virtual libraries into characterized new chemical substances, register them then move them rapidly into HTS for biological evaluation. The gap-spanning process is multicomponent and involves both new types of process control software combined with the emerging new device technologies. It also involves changes in established procedures to accommodate the high volumes of samples and information.

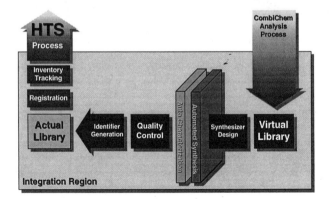

Figure 2. Component detail on the integration region.

The process can be discussed in terms of primary component function, for which there are three: automated synthesis, automated QC, and high throughput registration. Automated synthesis involves those steps required to transform the output from combinatorial chemistry analysis, the virtual molecular library, into fully populated microplates. Automated QC are the steps required to characterize the samples produced by automated synthesis. Characterization is the critical step in ensuring that the entire process produces meaningful results. Finally, high throughput registration assigns identifiers to samples and plates and updates the inventory systems.

Although easy to discuss these as separate components, the fact is unprecedented levels of integration between both the primary components discussed here, combinatorial chemistry software, and HTS software will be required to achieve and sustain both high sample volumes while preserving quality.

The risk of a new process is the introduction of new performance bottlenecks that result in lower than expected sample and data throughput. Each primary component has a bottleneck risk (Fig. 3) resulting from how the process control software handles situations that it cannot resolve. Each situation as it arises causes normal processing to stop and sample and information flow to halt while the situation is dealt with.

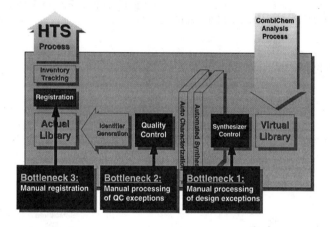

Figure 3. Performance bottlenecks introduced by the new process.

An unresolvable situation is referred to as an exception condition, or more simply, as an exception. It is not an error, although it is easy to interpret them as such, but rather a failure in decision process due to insufficient information, confused data, or inadequate program logic. Note that exceptions can occur in he reasoning processes of both software and humans.

The main source of process bottlenecking is manual intervention to support the inability of the process control software to handle all situations. It is unreasonable to expect that any software will be able to handle all exceptions. It is also inappropriate for the problems to be ignored. Manual intervention must be part of key decision steps in order to avoid excluding library members from synthesis and mischaracterizing the samples and keeping the quality of the overall research process high.

The goal is to limit and focus manual intervention as much as possible. The process control software must have the ability to reassess the exceptions using different rules so that the remaining exceptions are ones that should be passed to the chemist rather than simply just passed on. An example that will be elaborated on later is the controlling software for the automated characterization component. If it cannot successfully characterize the expected sample with the analytical information, it must try to determine what the sample might be or that no characterization is possible (e.g., mixture), and only on failing with any characterization pass the exception on for human intervention. This results in exceptions that are both focused and minimized as well as more appropriate for the skills of the chemist.

Other sources of process bottlenecking exist. Instrumentation time lag during QC is a potential bottleneck. This is not an exception condition as such, rather it is a true bottleneck. This is the time between automated QC data gathering and the actual characterization, and is instrument dependent. The time lag is lengthened in operations where multiple instruments are used, each with its own sample preparation and time requirements.

Exception conditions will always happen. The issue is minimizing the exception conditions to reduce the impact of these process flow diversions on overall sample throughput without sacrificing accuracy. In practical terms, this means that at each stage the controlling software must be able to handle a broader range of problems, adapt to new information, and limit the exceptions put out for resolution.

Synthetic Planning

The mission of automated synthesis is to transform a virtual library of samples into fully populated plates of samples. In principle, the virtual library could come from any source, but more often will come from combinatorial chemistry design software. The library consists of those samples that have been committed to synthesis. Full plates make for more efficient HTS. The process is shown in Figure 4.

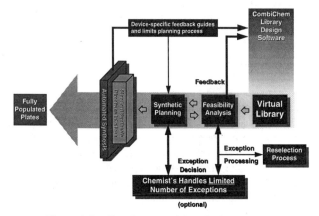

Figure 4. Detail on the automated synthesis component.

The process control software consists of two stages, discussed here for convenience, but more likely will be combined into a single application. The first stage assesses the feasibility of the desired molecule with respect to synthesizer capabilities. The second stage deals with the details of chemical synthesis and sends that information on to the automated synthesizer. The two stages work together because feasible may not always translate to doable. The result is the molecules and directives that are passed on to the automated synthesizer.

Exceptions occur in both stages. Molecules that are clearly not makeable can be overcome by adding a library reselection process that selects a replacement member for the one that failed. This works if the library being made is a subset of the actual library and all members were prioritized. Reselection involves choosing a lower priority molecule from the set that was not originally chosen. Otherwise, the option to reselect has to go back to the combinatorial chemistry software for sources of replacements or to fall through to the chemist for resolution that might include either replacing or skipping the exception.

Exception handling for the synthetic planning stage will be constrained by the limitations of the synthesizer technology. A possible but impractical choice is to manually synthesizer the exception molecule that would introduce a serious bottleneck. More likely choices will be adjusting synthesizer parameters within the scope of the synthesizer's capabilities.

Feedback will be important for enhancing the knowledge base of several components in this process. The process is learning while it operates and capturing that information and applying it to later operations will save time and improve operating efficiency. One important feedback pathway is feasibility information passed back to the combinatorial chemistry software for later use. Most likely this will involve updating a shared database.

Synthesizer capability feedback is another potentially valuable information exchange mechanism in which the synthesizer dynamically imposes limitations on upstream decision making. Were this to become a standard, the addition or update of synthesizer capabilities would not require reprogramming of the control software to take into account new devices. Coupled with a standard interface to the synthesizer, then both the control software and automated synthesizer become modular components that benefits both users and developers of this technology.

Automated Quality Control

The mission of automated quality control is to establish the nature of the substances in each well. Although quality control acceptance criteria are organization-specific, for libraries of discrete small molecules, characterization must be sufficient to establish both chemical structure and purity. At issue here is that synthesizer failure may produce a pure, but unexpected compound, so the question answered by automated QC is "what is it" rather than "is it the expected sample?"

Automated QC is accomplished in a three-step procedure: (1) characterization data acquisition, (2) data interpretation, and finally (3) data assessment, as shown in the process arrow of Figure 5. The goal is to characterize all samples on a plate that includes flagging the sample as unusable, so that downstream analysis of HTS screening data with the substances can be done in a reliable manner with diminished chance of false positives caused by improper follow-up of incorrect structures.

For the data acquisition component, there are instruments today that are capable of performing high throughput data generation on samples already stored in plates. A good examples of this are the readers used to generate robotics assay data. At issue is that the range of data types, such as NMR techniques, are inadequate to provide a general method for automated small molecule characterization. This is a serious technological gap and one that is an opportunity for instrumentation developers.

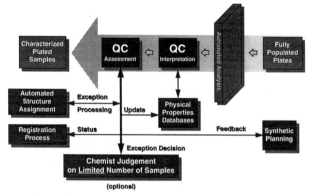

Figure 5. Detail on the automated quality control component.

Data interpretation judges whether the data acquired from the instrumentation agrees with the expected structure assigned to that well from the upstream library management software. Purity must also be part of the interpretation. The result is a yes/no answer: it is or it is not the expected structure. Most of this is established technology applied to high throughput situations with the new technology consisting of the integration of data and interpretations from multiple source devices.

Data assessment is, in effect, a decision process manager that manages a cascade of responses based on the initial characterization. The principal response is a reassessment process that helps leverage existing information and limit the number of exceptions required to be processed by a chemist. Other responses include updating the characterization database to reflect information gained during the assessment process, preparation of information for the downstream registration process, and feedback to the upstream synthetic planning software. Communication with the registration process allows the status of characterization to be included in registration as well as passed further downstream for use in the interpretation of screening data. More will be said about this in a later section. Communication with the synthetic planning software provides feedback to alter future planning choices and let the process take advantage of information gained.

The process must support a second round of evaluations to attempt to determine what the sample is once it has been determined that it is not what it should be (Fig. 6). At issue is characterizing all plated samples because all will be screened in the downstream HTS process and minimizing the bottleneck that results when exceptions require manual characterization. A requirement of manual exception handling will be electronic handling of analytical information to speed the overall process. There is also ample opportunity for pattern generation and recognition technologies to be developed for this problem.

The assessment process assigns a structure or provides a situation-dependent rejection response. Because the samples are already plated and will advance into screening, rejection can be classified as either UNKNOWN or VOID. The UNKNOWN label passed downstream marks screening hits as requiring further characterization before the hit's biological activity can be assessed. This takes advantage of the presence of the sample even if it remains unknown. The other choice is to VOID the sample and treat it as if it were an empty well regardless of data generated. Of the two choices, the UNKNOWN label provides the greatest flexibility for the future.

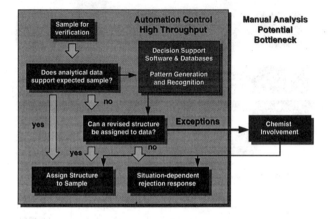

Figure 6. Sample characterization decision logic that minimizes manual analysis.

The assessment process can narrow and focus the exceptions by using a three-tier approach to automated decision making. This approach accepts correct assignments, rejects mixtures and impure samples that cannot be assigned without exception generation, and passes only those exceptions that could be characterized to the chemist. This takes good advantage of the chemist's skills and time while limiting the number of exceptions. Again, this additional level of flexibility in the decision process creates new opportunities for analysis software.

Automated QC and HTS can be run in parallel up to the point where screening data is analyzed (Fig. 7). Automated QC will require time for data generation and acquisition, especially in an environment consisting of multiple instruments. The process of screening also requires time and does not require characterized samples. An automated process that can update the characterization status of samples as they move through the process allows the QC data to catch up with the screening and prevent another bottleneck. This status may range from validation to structural reassignment to assignment as UNKNOWN.

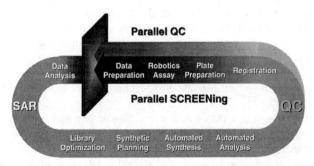

Figure 7. Parallel Screening and QC and their decision convergence point.

Although characterization is critical for establishing SAR patterns and is a requirement for analyzing screening data, the fact is that only the actives require adequate characterization. Due to the high sample and data volumes expected for the overall process, the analyst must necessarily focus on the actives. Most inactives are not examined except on a case-by-case basis. This is a practical concern because properly conducted HTS assays primarily produce inactives. Characterization should not be avoided for the inactives because multiple assay environments may turn up a hit from a previously inactive sample.

High Throughput Registration

The final component of the process is the registration of samples and plates to realize the library of actual samples ready for deployment into HTS (Fig. 8). Registration assigns official identifiers to the samples that tie the screening data to the sample. Master plates that act as primary inventory storage for sample libraries must also be registered and tracked.

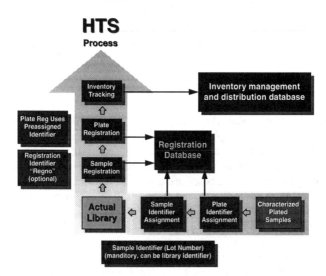

Figure 8. Detail on the high throughput registration component.

High throughput registration requires automated assignment of identifiers and coordination with the registration database in order to avoid another process bottleneck. Samples and plates require a source-level identifier and a compound and library identifier, in essence a lot number and a registration number when applied to the sample.

Most organizations require a chemical structure for the sample as a prerequisite for registration. A strategy for high throughput registration is to assign only sample-level identifiers, in short lot numbers, to samples. As samples are adequately characterized, the registration number can be assigned. This strategy is compatible with the parallel characterization and screening strategy of Fig. 7.

The Future

The future potential of automated synthesizers will likely be realized through multiple specialized devices that together cover the range of synthetic planning and execution required by an organization. A strategy is shown in Figure 9 in which multiple synthesizers are managed by a single synthesis planning software package with the output directed onto plates with the aid of an inventory robot. This flexible approach expands the range of targets accessible by the automation technology.

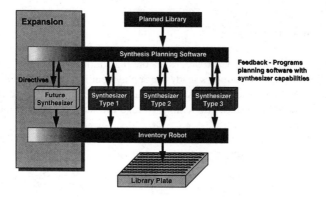

Figure 9. Flexible synthesizer capability and expansion strategy.

Feedback between devices is critical for this process to adapt and evolve in a modern research environment. As discussed earlier, feedback from synthesizers to planning software allows the planning software to be, in effect, dynamically programmed with real synthesis capabilities. The planning software can know which synthesizers are on-line and their range of capabilities. The capabilities can also include information on which syntheses are efficient and which are not. The planning software can exploit this information to route samples through the synthesizer matrix to optimize the production of the desired library.

This strategy also allows for adaptation for the future synthesizer technology using a plug-and-play approach successfully used for computer peripherals. The new synthesizer programs the planning software and becomes just another exploitable resource. The advantage of this approach is in the potential speed of deployment in bypassing the programming step. Of course, the interface technology will have to be robust to accommodate new types of instructions not envisioned in the original design.

Summary

A process for bridging between combinatorial chemistry and high throughput screening that exploits the emerging automated synthesis technology was described that has the potential for achieving high sample and data volumes expected for these enabling technologies. Bottlenecks in the proposed process were identified along with approaches for diminishing their impact by intelligent analysis and process decisions as implemented in new software. New technologies such as automated QA and its analysis software were identified as opportunities.

A key component of the process is information feedback that exploits the information gained from automated synthesis. Feedback from synthesizers has the potential for providing a simplified means of expanding and adapting new synthesizer technologies as they emerge. Multiple synthesizers coordinated by a single planning software system has the potential to dramatically expand the range of synthetic capability available to the organization.

The vision presented does not exist today. It is consistent with the requirements of large sample and data management that is a key component of the future of bioagent discovery envisioned by many organizations. Much new technology will be required for both individual components as well as to achieve the overall integration to create a workable process to span the gap. The purpose of this paper was to define the requirements and present the issues and opportunities for new technologies to truly enable high throughput discovery.

ANALYTICAL METHODS AND SCREENING

Chapter 22

Tools for Combinatorial Chemistry

**Keith Russell[1,3], Derek C. Cole[1], Frances M. McLaren[1],
and Don E. Pivonka[2,3]**

[1]Medicinal Chemistry Department and [2]Lead Discovery Department,
Zeneca Pharmaceuticals, Wilmington, DE 19897

A simple, fast, and non-destructive analytical methodology is described for the quantitative analysis of solid phase reactions without the uncertainties associated with cleavage of the compound from the polystyrene resin. The technique is based on Fourier transform infrared analysis and quantitation of deuterium labeled protecting groups.

The emergence of compound library synthesis as a potentially highly productive tool in the search for new therapeutic agents is revolutionizing medicinal chemistry. The parallel synthesis of *many* compounds, as opposed to serial synthesis of *single* compounds, has been driven by high throughput biological screens and their insatiable hunger for diverse compound sets.[1-3]

Drug Discovery Using Compound Libraries.

There are three key elements provided by a compound library-based drug discovery effort. These elements should lead to increased feasibility of finding novel leads and quickly optimizing known leads. The first element is "the numbers game" or the belief that the more compounds that are screened the more likely it is that a "hit" will be found based on simple chance.

A second key element to evolve is the introduction of a rational design or bias library. Given the huge number of potential drug-like small molecules with a molecular weight of less than 500, for example, it is unreasonable to ever assume comprehensive representation with a "random" library. Even relatively large 10^6

[3]Corresponding authors

1054–7487/96/0246$15.00/0
© 1996 American Chemical Society

compound libraries, designed with attention to molecular diversity, yield only a very small fraction of possible compounds. An example of a biased library would be a structure-based library developed around X-ray diffraction data of a ligand/protein complex, or based around an important protein motif. It is believed that a *marriage* of the above two elements is an optimal strategy for efficient drug discovery and optimization.

A third element of great importance is the "diversity" inherent to the compound library being generated. While important, it is currently difficult to adequately define what is meant by molecular diversity. This is probably a consequence of our inability to fully understand the details of the molecular recognition process itself. The protein target is, of course, the final judge of what constitutes a significant difference between two molecules in terms of their respective binding energies. It should be emphasized that the compound library approach is one of several tools available to the medicinal chemist, and does not replace "traditional " medicinal chemistry, which will always be important.

Compound Library Synthesis - The Choices. We have tended to subdivide compound library generation into two major categories - combinatorial chemistry, dealing with compound mixture synthesis, and multiple parallel synthesis (MPS), the parallel synthesis of many single compounds. These are, of course, two extremes which overlap in many cases. It should be recognized, however, that the term combinatorial chemistry is often viewed as a name for the whole field of compound library chemistry.

Solid Phase Synthesis - Benefits and Problems. Two strategies for the generation of compound libraries are common - solution phase synthesis and solid phase synthesis. Due to the benefits and limitations associated with each approach, the best approach varies with the design of the library and the requirements of the biological screen. Solid phase synthesis of small molecules has recently received much attention due to the ease of which tethered compounds can be separated from large amounts of excess reagents, impurities, and etc. While there are over three decades of experience in the area of solid phase peptide synthesis, by comparison, solid phase synthesis of small molecules is still in its infancy. This field is currently exploding in terms of the number of publications appearing in the literature.

One of the major problems associated with deploying solid phase synthesis is the lack of analytical methodology to support the synthetic effort. This is largely due to the fact that the analytical methods used for solution phase chemistry are often ill-suited for analysis of reactions on the solid phase.

Monitoring Solid Phase Synthesis. Several techniques can be used to study solid phase reactions including the following:

- Microanalysis
- Nuclear Magnetic Resonance Spectroscopy (NMR) [4,5]
- Cleave and Measure
- Other Methods
- Infrared Spectroscopy (IR) [6,7]

Microanalysis is a destructive technique commonly used to monitor changes in halogens, sulfur, and etc., in solid phase synthesis. Unfortunately, the C and H contribution of styrene backbone preempt analysis of these elements in solid phase synthesis. Changes in nitrogen content are small and therefore often difficult to detect. Routine NMR is of limited value for monitoring solid phase reactions and characterizing products. However, magic angle spinning NMR techniques increase the utility of solid/gel phase analysis and are becoming widespread in labs performing solid phase chemistry. The cleave and measure approach is commonly used to avoid the intricacies of on-bead analysis for the evaluation of solid phase synthesis reactions. In this approach, product is cleaved from the resin using chemistry dictated by the choice of resin linker and characterized using traditional solution phase methods. One problem with this approach, however, is that the cleavage chemistry is a variable that may lead to an incorrect interpretation of what is happening on the solid phase. Although additional analytical methods (e.g. MS, etc.) exhibit some further utility, simple, fast, sensitive, and non-destructive techniques for monitoring solid phase reactions are lacking. Outlined below is a description of a very simple infrared spectroscopic tool to help fill this void.

Quantitative Infrared Spectroscopy on the Solid Phase Support.

We have recently implemented quantitative infrared spectroscopy as a tool for the on-bead analysis of solid phase reactions. Although organic solution phase infrared analysis has long been a mature science, the on-bead analysis of solid phase combinatorial libraries presents several new challenges which must be overcome if infrared spectroscopy is to be a viable way forward for these analyses. These hurdles include: bead size; sensitivity; distortion of the infrared beam path due to indices of refraction at the air/bead/air interfaces (lensing); the inherently poor selectivity of infrared for differentiation of similar functional groups, i.e., acid, ester, aldehyde; the localized matrix influence on the cross-sectional molar absorbance of a functional group in question; and ill-defined bead diameters, porosity, and inhomogeneity which result in sampling pathlength variations.

Bead size, sensitivity and index of refraction considerations were addressed using a NicPlan™ infrared microscope fitted with a 0.1 mm liquid nitrogen cooled mercury cadmium telluride detector. Discrete bead analysis was conducted by imbedding the bead in a KBr pellet support to provide more efficient index of refraction coupling of the infrared sample beam to the bead surface, see Figure 1. The infrared beam was then apertured to the center portion of the bead to minimize absorbance nonlinearities due to pathlength variations throughout the analytical cross-section. In this configuration, small bead size and infrared sensitivity issues were eliminated. Lensing of the infrared beam at the bead surface was still present but attenuated. Introduction of a novel deuterium labeled protecting group methodology as detailed throughout the remainder of this chapter was found to be a powerful solution to infrared selectivity and matrix/induction interference concerns.

Figure 1: KBr Mounted 70 μm Bead

Deuterium Tagged Compounds. There are three clear benefits of using deuterium tagged compounds for quantitative infrared studies. First, the C-D stretching mode absorbances occur in a frequency region (2300-2200 cm^{-1}) which is clear of most other functional group absorption bands. Secondly, many deuterated starting materials are commercially available. Finally, incorporation of deuterium does not significantly affect reaction rates relative to hydrogen.

Our first pilot experiments involved the simple benzoylation of an aminomethyl polystyrene resin. The goal of these experiments was to provide a method for quantifying reaction yield by *analyzing a single resin bead in less than one minute*. Although it is difficult to generate true calibration standards for reaction yields between 0% and 100%, substoichiometric yields can be easily and accurately simulated to permit generation of the regression calibration. Apparent yields of 0% 20%, 50%, 80%, and 100% were obtained by quantitatively mixing benzoyl chloride and d5-benzoyl chloride in ratios of 100:0, 80:20, 50:50, 20:80, 0:100, respectively. Quantitative reaction of the calibration standard mixtures were insured by monitoring the C-D absorbance over time for the d5-benzoylation reaction. Upon completion of the reaction, excess reagent was added at elevated temperature with no further reaction noted. A Kaiser test was also used to confirm that no free amine remained.

Figure 2 illustrates the absorbance spectra obtained for these standards. Infrared throughput variations due to scatter and/or lensing of the infrared beam are evident through the elevated baselines in non absorbing spectra regions. Bead to bead infrared pathlength variations are also illustrated by the varying intensity of the 2335 cm^{-1} polystyrene absorbance.

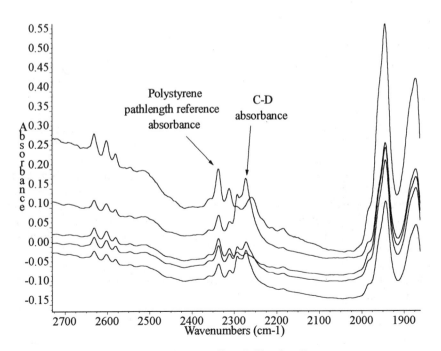

Figure 2: Amide Coupling Calibration Spectra

Figure 3 illustrates first derivative spectra calculated from the raw absorbance data presented in the Figure 2 illustration. Effective elimination of baseline slope and offset through the derivative process are apparent in Figure 3. In this figure, derivative absorbance intensity for each profile has also been normalized using the 2340 cm^{-1} backbone derivative absorbance to compensate for pathlength variations as a function of bead diameter and homogeneity.

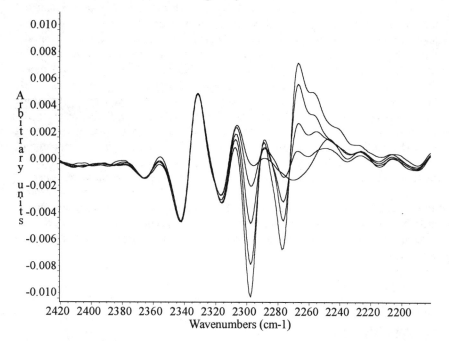

Figure 3: Pathlength Normalized Amide Coupling Derivative Spectra

Regression analysis of the deuterium to styrene derivative absorbance ratio vs. the simulated deuterium yield are presented in Figure 4. With an R^2 correlation coefficient of 0.9994, the precision of this calibration significantly exceeds the requirements of the experiment.

These experiments were also repeated for two other coupling reactions. Ester coupling was effected via benzoylation of a tethered alcohol. Acetamide coupling was obtained by reaction of aminomethyl resin with mixtures of acetic anhydride and its d3 counterpart. In each case, an excellent correlation was obtained between the simulated yield and the measured deuterium to styrene first derivative absorbance ratios. In the case of d3-acetylation, the aliphatic d3 absorbance was significantly weaker than for the aromatic d5-benzoylation experiment. However, signal to noise was more than adequate to render sample variations, as opposed to detector noise, the sensitivity limiting factor.

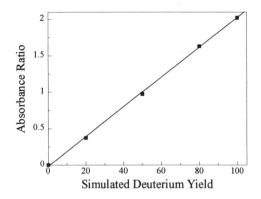

Figure 4: Amide Coupling Yield Derivative Absorbance Calibration

Environmental Dependence of the C-D Infrared Absorbance. In the case of the benzamide, benzoate ester, and acetamide reactions described above, both the frequency profile and molar absorbance of the first derivative C-D bands were internally consistent within a given functional group series. However, differences were noted between discrete functional group series. Hence, separate calibration curves needed to be generated for quantitation of each type of coupling reaction in question.

A more useful tagging protocol would ensue if the C-D absorbance profile and magnitude were independent of the matrix environment. Elimination of induction and matrix effects would allow construction of a single calibration for use in monitoring many different reactions. The possibility of a universal calibration was explored using deuterated protecting groups in which the CD functionality was chemically isolated from coupling site induction. One approach to the universal calibration is illustrated below with a common amine protecting group, the d9-tertiary butyloxycarbonyl (d9-tBoc) urethane moiety. This unit is often used as a protecting group of side chain amines in amino acids. Although, the example detailed here is based on a peptide, the d9-tBoc protecting group technology is equally applicable to small molecules.

Preparation and Utility of d9-tBoc Protected Amines. Commercially available perdeuterated d9-tertiary butanol was converted into the d9-tBoc-ON reagent using analogous methods to that used for preparing the non deuterated material. Fmoc lysine was then protected as its d9-tBoc derivative in the presence of a triethylamine.

Aminomethyl polystyrene was reacted with Fmoc glycine followed by deprotection with piperidine to expose the free alpha amino group. This procedure was further repeated to generate a triglycine spacer unit with a terminal free amine. Five distinct solid phase compounds were generated from this material, each one containing one d9-tBoc in a different environment according to Scheme 1.

Scheme 1.

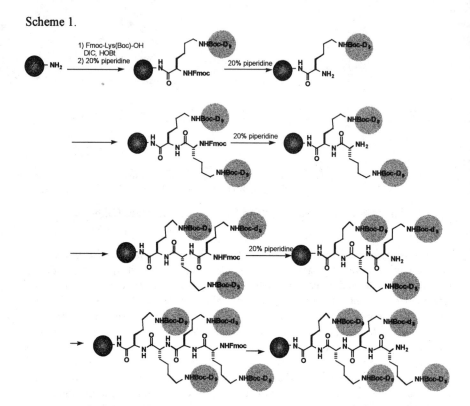

Analysis of washed beads from these experiments revealed that the profile and magnitude of the first derivative C-D bands did not significantly differ between the five compounds. These results indicated the d9-tBoc tether effectively isolated the C-D stretching absorbance of the d9-tBoc protecting group from perturbation by the parent molecule, within the scope of these experiments.

The success of these peptide experiments suggests the reactivity of tethered small molecules could analogously be determined using this technique. Furthermore, if d9-tBoc were incorporated into the substituent to be added, the analysis would not require a new calibration to be generated for each new reaction. Similarly, deprotection steps or the cleavage of the final product from the resin could be explored by monitoring the disappearance of the d9-tBoc absorbance. No detailed structural information is provided by this technique such as can be provided by direct analysis of the tethered compound either by "traditional" infrared or NMR, however, the speed, selectivity and quantitation inherent to the deuterium tagging methodology is unmatched by either of the aforementioned direct analysis techniques. It is anticipated that this method can be used in much the same way as thin layer chromatography is used for simple solution phase chemistry systems, in that it is a fast and simple way to monitor reactions.

Counting Tethered Lysines. The power of this technology can be further illustrated with resins containing one to four d9-tBoc lysine units per tethered molecule. In practice, four pairs of compounds were generated. Each pair consisted of the Fmoc protected and the deprotected free alpha amino terminus unit. In this experiment, pathlength normalization was obtained by the ratio of the CD derivative absorbance at 2222.2 cm^{-1} to the 2596.6 cm^{-1} - 2601.8 cm^{-1} styrene derivative peak to baseline value. In Figure 5, the absorbance ratios for compounds containing one through four d9-tBoc units are plotted against an extrapolation of the regression calibration line generated in the previous section (Preparation and Utility of d9-tBoc Protected Amines). Absorbance ratio data for the one through four d9-tBoc lysines (◊) overlay the extrapolated regression line so closely that the trace appears as a regression of the one through four d9-tBoc lysines although it actually is not. As expected, the presence of Fmoc vs. free amino terminus materials introduced no interference into the D9 quantitation. Hence, d9-Boc protected lysines are able to be counted in the solid phase sample.

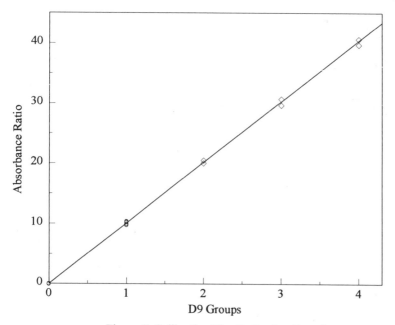

Figure 5: Calibration Plot for Lysine Counting

Possible Applications

Infrared analysis of single resin beads containing deuterium tagged compounds has been demonstrated to be a fast and quantitative method for monitoring solid phase reactions. The method eliminates any uncertainty associated with the cleavage of the compound of interest from the resin. Reactions amenable to this technology include bond formation, bond cleavage, deprotection etc. provided a deuterium labeled protecting group can be incorporated into one of the components being added or cleaved.

Conclusions

- Quantitative FT infra-red spectroscopy of single resin beads can be performed in less than one minute using commercially available equipment and is clearly a useful tool in solid phase synthesis and combinatorial chemistry.

- Incorporation of deuterium via deuterated protecting groups or otherwise can greatly simplify the quantitative analysis, while not significantly perturbing reaction rates of the parent hydrogen containing compound.

References

[1] Doyle, P. M., J. *Chem. Tech.* **1995**, 64, 317-324.

[2] Dewit, S. H.; and Czarnik, A. W., *Curr. Opin. B.* **1995**, 6, 640-645.

[3] Chabala, J. C., *Curr. Opin. B.* **1995**, 6, 632-639.

[4] Blossey, E. C.; and Cannon, R. G., *J. Org. Chem.* **1990**, 55, 4664-4668.

[5] Look, G. C.; Holmes, C. P.; Chin, J. P.; and Gallop, M. A., *J. Org. Chem.*, **1994**, 59, 7588-7590.

[6] Yan, B.; Kumaravel, G.; Anjaria, H.; Wu, A.; Petter, R. C.; Jewell, C. R. Jr.; and Wareing, J. R., *J. Org. Chem.*, **1995**, 60, 5736-5738.

[7] Yan, B.; Kumaravel, G., *Tetrahedron*, **1996**, 55, 843-848.

Chapter 23

Magic-Angle Spinning NMR Spectroscopy of Polystyrene-Bound Organic Molecules

Monitoring a Three-Step Solid-Phase Synthesis Involving a Heck Reaction

Christophe Dhalluin, Iuliana Pop, Benoit Déprez, Patricia Melnyk, André Tartar, and Guy Lippens

Chimie des Biomolécules, Unité Recherche Associée, Centre National de la Recherche Scientifique 1309, Institut Pasteur de Lille, 1 rue du Pr Calmette, 59019 Lille Cédex, France

Organic synthesis by solid phase methods is emerging as a powerful tool for the combinatorial synthesis of large arrays of small molecules. The major difficulty in organic solid phase synthesis development lies in the structure determination of the compounds still tethered to the insoluble matrix. Until now, cleavage of the tethered molecule from the support and classical homogeneous phase methods were employed for the control of products synthesized on solid supports. Recently, the possibility of performing NMR analysis of polymer-linked molecules has been reported. We report here the MAS NMR gel phase monitoring of a three-step synthesis performed on a divinylbenzene crosslinked polystyrene. Theoretical and experimental aspects of the MAS NMR technique applied to organic solid phase synthesis are discussed.

High-resolution Magic Angle Spinning (MAS)[1,2] NMR has recently attracted considerable attention as a method to analyze the products resulting from organic solid-phase synthesis (SPS) without the need of cleaving them from their solid support.[3-5] Whereas the NMR assignment of compounds still attached to the insoluble matrix using the standard high resolution techniques remains a real challenge due to the line broadening caused by the restricted molecular motion of the molecules on the polymer beads and by the bulk magnetic susceptibility discontinuities in the rotor, the use of the MAS technique has allowed to overcome some of these limitations. In this report, we explore a number of experimental factors and their influence upon the quality of the NMR spectra, we discuss the general assignment strategy of a compound on a bead and exploit the previous results to demonstrate the step-by-step assignment of a chemical reaction without the cleavage of the intermediate products.

1054–7487/96/0255$15.00/0

The solvent motion in the polymer network leads to chain expansion observable on a macroscopic scale as a swelling of the bead and is accompanied by an enhanced local mobility of the attached molecules at the microscopic level. As a consequence of the enhanced molecular mobility the various interactions that govern the line width of the NMR signal will be motionally averaged, including the homonuclear dipolar interaction that can broaden proton signals up to values of 20 kHz in a static solid state sample. This motional averaging may be incomplete, due to the anisotropic environment of the solvent molecules in the polymer. The resulting residual dipolar broadening can be further reduced by spinning the sample at the magic angle at a spinning rate higher than the non-spinning residual line width. This last condition on the spinning rate can be advantageously exploited to suppress the lines of the more rigid supporting polymer while obtaining a liquid-like spectrum for the more mobile attached molecules.

MAS will also eliminate the magnetic-susceptibility broadening caused by the difference in (electronic) magnetic susceptibility between the polymer and the pure solvent, combined with the irregular shape of the solvent/polymer interface.[3-5] Although removed by the same experimental operation of spinning the sample at relatively low speeds (a few kHz), the two phenomena of dipolar broadening and magnetic-susceptibility broadening should be clearly distinguished. This distinction was made very early in a pioneering study on the NMR behaviour of hydrogen trapped in NbH_x.[6] The hydrogen diffusion is rapid enough to remove any homonuclear and heteronuclear dipolar broadening, but static linewidth for the proton signal as large as 13 kHz was observed.[6] Still, this static line width was accompanied by T_1 and $T_{1\rho}$ values superior to 10 ms, indicating that the broadening was entirely due to spatial inhomogeneities. The authors convincingly showed that it could be considerably narrowed down by spinning the sample at a rate largely inferior to 13 kHz.

We will show that in the case of MAS NMR on the support-attached products of solid-phase synthesis a spinning rate of 2 kHz, easily obtainable on a standard solid state probe, is sufficient to eliminate to a large extent both mechanisms of line broadening mentioned above, at the condition that certain experimental parameters of solvent and bead structure are obeyed. The influence of solvent and swollen bead structure with its linker on the resulting spectra can be understood in the framework of the motional averaging as described above. Observation of these simple rules leads to 1D and 2D homonuclear and heteronuclear spectra allowing the complete attribution of structurally complex molecules.

NMR Technical Aspects.

Whereas the first results of MAS NMR in SPS were presented using a specially designed probe (the 'Nano-probe', Varian NMR Instruments, Palo Alto)[3,7], it readily became clear that completely satisfactory results could be obtained on a standard MAS probe.[4,5] In the present work, we continue along the last lines : all our results were obtained on a Bruker 300 MHz DRX spectrometer, equipped with a pneumatic unit and a standard 7 mm MAS probe.

In high resolution NMR spectroscopy, the homogeneity of the static magnetic field is of paramount importance for the quality of the lineshape in a liquid sample, and is obtained by the well-known shimming procedure.[8] In solid state NMR, however, the intrinsic line broadening is such that extensive shimming is normally neglected. For the applications described here, it is clear that shimming - though different from the procedure used on an axially symmetric liquid sample - will be of uttermost importance. Indeed, whereas the local contribution to the field inhomogeneity (due to the many irregular surfaces) will be averaged out by MAS, the macroscopic field inhomogeneity (due to probe design, the various interfaces rotor/sample,...) should be minimized by the shimming procedure.

In high-resolution liquid NMR, the level of field homogeneity is usually evaluated from the intensity of the field lock dispersion curve.[9] The conventional 7 mm MAS solid state probe we used possesses no dedicated lock channel, but is equipped with two detection channels for the ^{1}H and X nuclei. For the initial shimming, the X channel was set to the ^{2}H nucleus frequency, and the probe was properly tuned in ^{2}H, allowing ready use of this coil for a field frequency lock. The adjustment of the shim coils has been performed by using the magnitude of the field frequency lock signal. This homology with the situation normally encountered in liquid high-resolution NMR facilitated the shimming protocol, and proved to be easier than shimming on the decay of the envelope of the *Free Induction Decay* signal.

The lack of axial symmetry with respect to the main static field, and the unfavorable ratio of sample length/diameter leads to deviations from the habitual shimming protocol.[10,11] The field inhomogeneities in the sample were compensated primarily by a fine adjustment of the radial shim coils X, Y, XY, (X^2-Y^2), X^3 and Y^3, and after that by a minor contributions of the Z coil and its higher order. A well symmetric signal with a final line width of 5 Hz could be obtained on the ^{1}H water resonance of a static sample of water (90%H$_2$O/10%D$_2$O) oriented at the magic angle.

The frequency lock as implemented through the X coil was also used in homonuclear spectra to minimize drifts in the magnetic field during the data collection. For heteronuclear spectra, however, the X coil was set to the heteronucleus (^{13}C), and these spectra were run without field locking. Still, field drifting was not important during the reasonably short duration of the HMQC spectra (3 hours), leading to high quality heteronuclear correlated spectra on the same standard MAS probe.

A spinning rate of 2 kHz was sufficient to obtain line narrowing and to avoid any spinning side bands in the ^{1}H spectra. For larger rates superior to 3 kHz, the rotation of the sample was harder to stabilize, and for some solvents the tightness of the rotor proved to be insufficient, leading to the escape of some solvent from the rotor. However, for samples where the higher spinning rate could be used, no significant improvement of the spectral quality was observed. Therefore, all further studies were performed at a 2 kHz spinning rate. Whereas the previous considerations are mainly concerned with the removal of as many as possible magnetic field inhomogeneities, a second important factor proved to be the residual dipolar coupling experienced by the attached molecules. This parameter is directly linked to the mobility of the molecules in the heterogeneous environment that are the swollen beads. It is not straightforward to evaluate this parameter, but in the next paragraph, we try to establish a correlation with the solvent-dependent swelling of the beads.

Physicochemical Characterization of the Resin Beads.

The preparation of the resin sample which has to be submitted to a MAS NMR experiment is extremely important to obtain workable data. One key issue is to find out the best swelling conditions for the resin sample. First, it is important to determine the minimal volume of solvent required to swell a given amount of resin. Indeed, a resin sample prepared for a MAS NMR experiment should macroscopically behave as a stable suspension of beads in the solvent. The bead packing occuring in the spinning rotor must not lead to the separation of two distinct phases as this would induce a deterioration of the magnetic field homogeneity. On the other hand, in terms of microscopic behaviour, as previously discussed, the mobility of the molecules tethered to the polymer is a determining factor for the resulting line width. This mobility depends itself on the solvation volume of the resin. We therefore tested different solvents on their ability to swell the resin, and set out to establish a correlation between this parameter and the resulting spectra.

Table 1. Solvation of the Wang-Lys(Boc)-Fmoc resin in different solvents.

entry	solvent	solvation volume [a] (μL/100 mg)	volume of swollen beads (μL/100 mg)	averaged diameter (μm)	volume ratio [b]
1	pyridine	700	1000	109.6	6.3
2	DCM	700	800	107.5	5.9
3	chloroform	750	900	103.4	5.4
4	benzene	620	900	102.9	5.2
5	DMF	540	900	92.7	3.8
6	DMSO	200	300	85.1	2.9

[a] minimal volume of solvent required to swell 100 mg of dry resin. [b] the ratio of solvated volume and the volume of dry resin as calculated from the averaged diameters

The resin which was further used for the MAS NMR experiments was a commercially available 1 % divinylbenzene crosslinked polystyrene Wang-Lys(Boc)-Fmoc resin. The minimal volume of solvent required to swell 100 mg of resin was determined. 50 μL portions of solvent were added to a known quantity of resin until solvent saturation of beads (appearing of a liquid phase). The volume occupied by the resin beads swollen in the different solvents was measured using a graduated test tube. The swelling ability of different solvents, associated with the volume ratio (ratio of a solvated bead volume to the volume of the dry bead) was then established more precisely. The averaged diameter of the dry bead was 59.5 μm.The polymeric bead volumes were determined by direct measurement of their diameters under a light microscope provided with a micrometric ocular. For the solvated resin, beads (200 - 400 mesh) preswollen in the solvent were spread over a slide with a coverglass to minimize solvent evaporation. The resin beads proved to have even diameters so a total of about 100 beads for each determination was enough to determine an averaged diameter. For all these studies, six solvents were tested (Table 1).

The volume of solvent necessary to completely solvate the beads correlates well with the final volume occupied by the individual beads. A very low amount of dimethylsulfoxide (100 μL) was sufficient to completely solvate the beads, and the final volume ratio shows that dimethylsulfoxide (DMSO) does not efficiently penetrate into the polymeric network. The same rule was observed when comparing the solvation volumes for the other solvents, with pyridine needing a maximal solvation volume and showing the highest volume ratio. However, when we regard the volume of solvent needed to saturate the beads with respect to the final macroscopic volume of the swollen beads (and not with the individual bead volume), the correlation is not obvious. Indeed, we observe an almost constant volume of the swollen beads (about 900 μL/100 mg resin), despite the different solvation volume used. This would indicate that larger swollen beads (eg. for pyridine) pack more closely than the smaller ones (eg. for N, N - dimethylformamide), leading to a similar final volume. Although we presently do not understand very well the underlying phenomenon, the resulting NMR spectra (see below) show very similar line widths, indicating that the mobility of the tethered molecules is not directly affected.

A good solvation of the bead is likely to be related to a wide relaxation of the polymeric framework and an improved mobility of the tethered molecules. The inability to obtain workable NMR data for the resin swollen in deuterated DMSO can be correlated to its poor swelling ability. The motional freedom for the internal molecules, which represent more than 99 % of the tethered molecules, is probably more restricted when using the DMSO than for the other solvents. Good resolution was obtained when the polymeric beads were swollen in deuterated pyridine, but spectra of comparable quality were obtained when working in deuterated for N, N - dimethylformamide (DMF) as well as in deuterated chloroform, dichloromethane (DCM) or benzene. There probably exists a threshold for the solvation volume above which the mobility of the tethered molecules becomes important enough to reduce the residual dipolar coupling.

The MAS ^{1}H spectra exposed sharp resonances with a line width of 21 Hz in deuterated chloroform, DCM, pyridine and DMF and of 32 Hz in deuterated benzene. The signals in the ^{13}C MAS spectra have an average line width of 13 Hz in each solvent. The high motional freedom exhibited by the protected lysine bound to the solvent swollen beads results in solution-like properties and hence in sharp resonances. These entirely satisfactory resolutions in ^{1}H and ^{13}C obtained with a conventional MAS probe are of similar quality to that obtained with a 'Nano' probe.[3]

Since swelling properties of a resin depends on the nature of the growing molecule fixed on the polymer -a fact well known to solid phase chemists- it is likely that a systematic study of the swelling should be performed after each step of the synthesis.

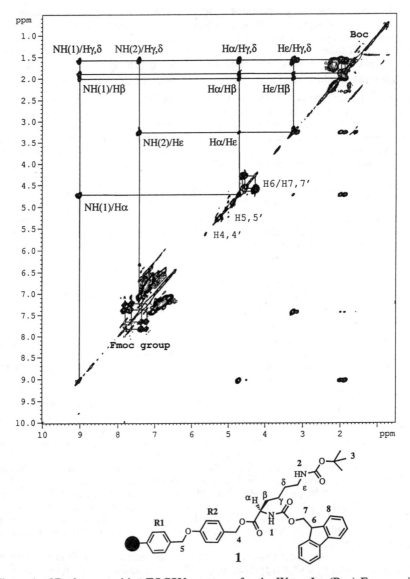

Figure 1. 2D phase sensitive TOCSY spectrum for the Wang-Lys(Boc)-Fmoc resin (1) swollen in the deuterated pyridine, obtained with a 70 ms MLEV-17 mixing sequence and with MAS at 2 kHz.

Attribution of a Molecule Attached to its Bead

Using a conventional 7 mm MAS solid state probe and the above described experimental conditions, 'solution' quality ^{1}H and ^{13}C NMR data could be collected for the Wang-Lys(Boc)-Fmoc resin (1). TMS was added as internal reference to the deuterated pyridine before the resin swelling, and the ^{2}H resonance of the solvent was used for the field frequency lock during the homonuclear data collection.

Although a line width of 21 Hz can be considered as broad for high resolution studies, this resolution proved to be entirely satisfactory for the extraction of structural information through two-dimensional NMR spectra. Total ^{1}H and ^{13}C NMR assignments and structural identification for the Wang-Lys(Boc)-Fmoc resin could readily be obtained using standard TOCSY[12] and ^{1}H-^{13}C HMQC[13] experiments. For the TOCSY experiment, a 70 ms MLEV-17 mixing time was used with a field strength of 10 kHz and appropriated delays to compensate for ROESY effects (clean-TOCSY).[12] The TOCSY and HMQC data were collected as a data matrix of 1024*256 complex points with 16 scans per increment. The total time for each 2D experiments was approximately 3 hours. These results for the deuterated pyridine swollen beads are shown in Figures 1 and 2, but results for the other solvents (except for deuterated DMSO) were comparable. The identification of the ^{1}H spin system of the protected lysine was unambiguous using the pattern of relayed connectivities in the TOCSY spectrum as described in the Figure 1. The chain can be traced from the backbone amide proton through the α-proton and the different side chain protons up to the ε-protons. The reverse pathway from the *tert*-butoxycarbonyl (Boc) protected amide proton through the ε-protons and all side chain protons until the α-proton can also be identified. The 9-fluorenylmethyloxycarbonyl (Fmoc) proton resonances are characterized by a typical pattern of connectivities in the aromatic region. The H6 and H7 protons connecting the Fmoc group to the Lysine residue were assigned as a group of three interconnected protons, whereas the H4 and H5 protons form isolated proton pairs and could not be unambiguously assigned at this stage.

The carbon assignments follow directly from the previously determined proton assignments and from the observed ^{1}H-^{13}C correlation peaks as shown in Figure 2. Complete ^{1}H and ^{13}C NMR assignments for the Wang-Lys(Boc)-Fmoc resin swollen in the deuterated pyridine are provided in the Table 2. The chemical shift assignment obtained here agrees well with the values obtained by Anderson and al. for the Wang-Lys(Boc)-Fmoc resin.[5]

Whereas the 2 kHz Magic Angle Spinning removes the residual static line broadening caused by the proton dipolar coupling, the dipolar cross-relaxation mechanism is unaffected by this process.[14] Therefore, a NOESY[15] experiment performed with a 600 ms mixing time yielded intense cross peaks between the protons spins of the attached molecules. Figure 3A shows the TOCSY spectrum for the Wang-Lys(Boc)-Fmoc resin swollen in deuterated benzene, which yields results very comparable to those obtained in deuterated pyridine. A part of the corresponding NOESY spectrum is shown in Figure 3B. NOE connectivities between the H6 and H7 proton resonances and the H8 proton resonance of the Fmoc group can be observed. The NOESY spectrum allows further an unambiguous identification of the H4 and H5 resonances. A sharp NOE peak was observed between H4 and two aromatic protons of the ring noted R2, whereas a broad NOE peak was observed between H5 and two aromatic protons of the ring noted R1 which already belongs to the polymer solid support. The line width read from these NOE connectivities is directly correlated to the motional freedom of the protons H4 and H5 in the molecule: H5 is closer to the polymer bead than H4, and presents a broader line width. The difference in line width observed in the ω_1 dimension clearly indicates the mobility gradient for the two aromatic rings R1 and R2: whereas the proton resonances of the R1 ring of the polystyrene bead are large, the presence of the ether linker confers an

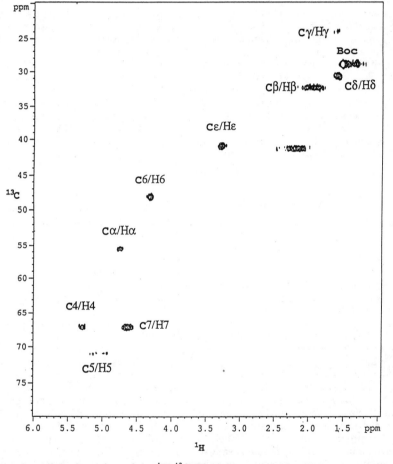

Figure 2. Aliphatic region of the ^{1}H-^{13}C HMQC spectrum for the Wang-Lys(Boc)-Fmoc resin (**1**) swollen in the deuterated pyridine with MAS at 2 kHz.

Table 2. Proton and Carbon-13 assignments for the Wang-Lys(Boc)-Fmoc resin (**1**) swollen in the deuterated pyridine.

Assignment	Proton shift	Carbon-13 shift
α	4.73	55.26
β	1.89 2.03	32.14
γ	1.59	23.85
δ	1.59	30.31
ε	3.26	41.13
NH(1)	9.03	-
NH(2)	7.43	-
Boc	1.53	28.73
4	5.27	66.80
5	5.00	70.46
Fmoc group aromatics	7.83, 7.65, 7.34, 7.24	nd
6	4.30	47.89
7	4.57, 4.66	66.80

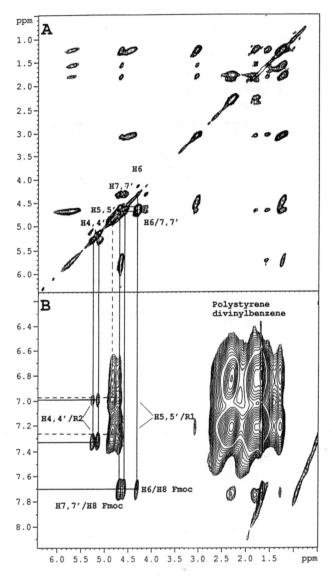

Figure 3. (A) 2D phase sensitive 70 ms TOCSY spectrum for the Wang-Lys(Boc)-Fmoc resin swollen in the deuterated benzene with MAS (2 kHz). (B) 2D phase sensitive NOESY spectrum for the Wang-Lys(Boc)-Fmoc resin (1) swollen in the deuterated benzene, obtained with 600 ms for mixing time under MAS at 2 kHz.

important degree of motional freedom to the R2 ring, resulting in a narrow line width for protons on the latter. In contrast to the TOCSY spectrum, intense and broad connectivities have also been identified between proton spins belonging to the polystyrene in the NOESY spectrum.

Step-by-Step Assignment of a Real-Life Chemical Reaction by MAS NMR.

A real challenge in solid phase synthesis is the monitoring of a multi-step chemical reaction. We decided to follow up the Heck reaction. This reaction is a well known process to form carbon-carbon bonds and presents a great interest for the combinatorial solid phase synthesis. It offers the possibility of exploiting a large variety of building blocks, generating disubstituted olefins which are of great interest in combinatorial chemistry. Moreover, the Heck reaction is generally a very mild process, thus being readily amenable to automation. Recently, the formation of carbon-carbon bond in solid phase synthesis using the Heck reaction was reported.[16,17]

The same Wang-Lys(Boc)-Fmoc resin of the previously described MAS NMR experiments was submitted to a three step synthesis (Scheme 1). First, the Fmoc group was deprotected in the classical manner, using a 20 % solution of piperidin in DMF. Secondly, the 3-iodobenzoic acid was coupled to the liberated αNH_2 group using PyBrOP as activating reagent. The completion of the reaction was determined by quantification of the residual NH_2 on the resin, using the quantitative ninhydrin procedure.[18] An aliquot of the resin was cleaved using a 90 % trifluoroacetic acid (TFA) solution in DCM and the product was analysed in HPLC. Finally, the aryl iodide on the resin was reacted with ethyl acrylate in DMF, in presence of tris(2-tolyl)phosphine, Palladium acetate and triethylamine, under nitrogen at 60 °C, to generate a disubstituted olefin. An aliquot of the final resin was also cleaved using a 90 % TFA solution in DCM and the released product was analysed in HPLC and in solution phase NMR. We thus confirmed the total convertion of the aryl iodide on the resin into the desired olefin.

Each reaction step was followed by MAS NMR TOCSY and HMQC spectra, using the experimental conditions previously determined for the Wang-Lys(Boc)-Fmoc resin. Pyridine was found to be the solvent of choice to swell the polymeric beads for the resins in all the three reaction steps (only TOCSY spectra collected using deuterated pyridine swollen beads are presented). At each reaction step, total structural identification and NMR assignment of the new molecular entity formed on the polymer could be unambigously obtained on the NMR spectra (Table 3, Figures 4 to 6).

For compound **2**, peaks corresponding to the Fmoc protecting group disappeared. In the same time, the shift of the α-proton toward the high field confirms the elimination of the Fmoc moiety which was present in the proximity of the $C\alpha$ position. The αNH_2 protons have disappeared from the spectrum. This inability to observe amino proton resonances in MAS NMR was previously reported in the case of the NH^+_3 of the L-alanine.[19] It was explained by an interference phenomenon between the molecular dynamics and the characteristic radiofrequency field strength of homonuclear pulse trains employed in the NMR experiments leading to dramatic broadening of lines of amino protons.

For compound **3**, peaks that appeared in the aromatic zone of the TOCSY NMR chart could be attributed to the 3-iodobenzoyl group newly attached to the resin. Only three from the four protons belonging to the 3-iodobenzoyl group were detected and we could not identify which peak corresponded to a defined proton. At this step, the important downfield shift noted for the α-proton confirms the presence of the benzoyl ring in the proximity of the $C\alpha$ position.

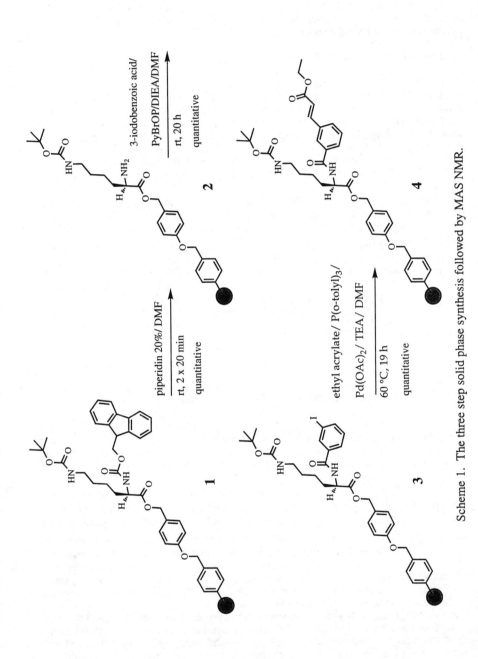

Scheme 1. The three step solid phase synthesis followed by MAS NMR.

Table 3. Proton assignments for resins **2**, **3** and **4** swollen in the deuterated pyridine.

Assignment	Proton Shift		
	resin **2**	resin **3**	resin **4**
α	3.55	5.11	5.14
β	1.80	1.94, 2.05	1.94, 2.06
γ	1.57	1.56	1.54
δ	1.57	1.56	1.54
ε	3.28	3.24	3.25
NH(1)	-	9.64	9.91
NH(2)	7.41	7.45	7.42
3	1.52	1.50	1.49
4	5.22	5.27	5.27
5	5.04	4.95	4.98
R3		8.63, 8.17, 7.76	8.47, 8.23, 7.60, 7.34
9			7.78
10			6.52
11			4.18
12			1.15

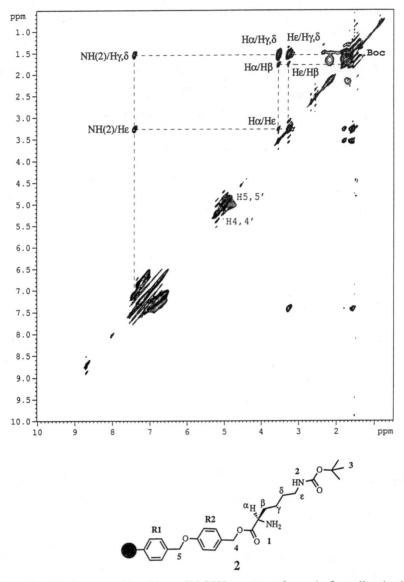

Figure 4. 2D phase sensitive 70 ms TOCSY spectrum for resin **2** swollen in the deuterated pyridine under MAS at 2 kHz.

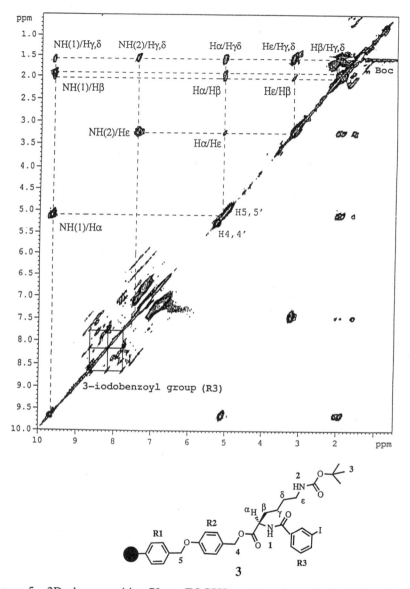

Figure 5. 2D phase sensitive 70 ms TOCSY spectrum for resin **3** swollen in the deuterated pyridine under MAS at 2 kHz.

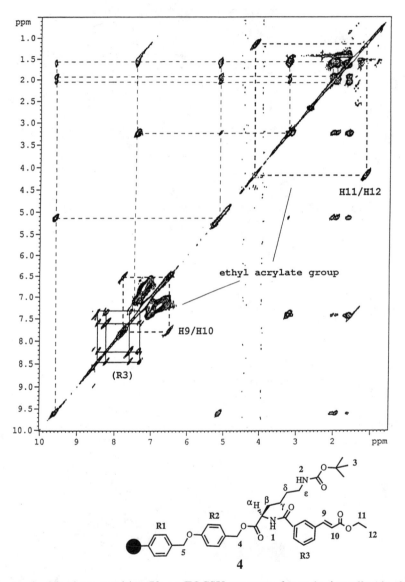

Figure 6. 2D phase sensitive 70 ms TOCSY spectrum for resin **4** swollen in the deuterated pyridine under MAS at 2 kHz.

In compound **4**, apparition on the TOCSY NMR chart of peaks attributed to the ethyl acrylate group confirmed the formation of the trans disubstituted olefin on resin, *via* the Heck reaction. This time, the four protons corresponding to the R3 aromatic ring are much better resolved than in compound **3** but once more we could not identify which peak corresponded to a defined proton.

Conclusions.

The MAS NMR technique was used to follow a real-life multi-step solid phase synthesis, without cleavage of the products from their solid support. At each reaction step, NMR assignment and total structural characterization of the new molecular entity formed on the polymer could be unambiguously obtained. A good swelling of the resin beads and the use of the MAS were found to be essential to obtain spectra of good quality. 'Solution' quality ^{1}H NMR data could be collected using a conventional 7 mm MAS solid state probe on a Bruker DRX-300 spectrometer.

Acknowledgements.

We thank Dr. P. Lux and Dr. M. Piotto (Bruker Spectrospin France) for the solid MAS probe, and Y. Dumazy (Université des Sciences et Technologies de Lille) for his help in obtaining these data. We are grateful to Prof. J. Jeener and Dr. P. Broekaert (ULB, Brussels, Belgium) for careful reading of the manuscript. C.D. acknowledges receipt of a graduate fellowship cofinanced by Bruker Spectrospin France, and the Région Nord-Pas de Calais (France). The 300 MHz facility used in this study was funded by the Région Nord-Pas de Calais (France), the CNRS and the Institut Pasteur de Lille.

References

1. Lowe, I.J. *Phys. Rev. Lett.* **1959**, *2*, 285
2. Andrew, E. R.; Bradbury, A.; Eades R. G. *Nature* **1958**, *182*, 1659
3. Fitch, W. L.; Detre, G.; Holmes, C. P.; Shooreley, J. N.; Keifer, P. A. *J. Org. Chem.* **1994**, *59*, 7955
4. Anderson, R. C.; Jarema, M. A.; Shapiro, M. J.; Stokes, J. P.; Ziliox, M. *J. Org. Chem.* **1995**, *60*, 2650
5. Anderson, R. C.; Stokes, J. P.; Shapiro, M. J. *Tetrahedron Lett.* **1995**, *36*, 5311
6. Stoll, M. E.; Majors, T.J. *Phys. Rev. B* **1981**, *24*, 2859
7. Manzi, A.; Salimath, P. V.; Spiros, R. C.; Keifer, P.A.; Freeze, H.H. *J. Biol. Chem.* **1995**, *270*, 9154
8. Fuks, L. F.; Huang, F. S. C.; Carter, C. M.; Edelstein, W. A.; Roemer, P. B. *J. Mag. Res.* **1992**, *100*, 229
9. Hoult, D. I.; Richards, R. E.; Styles, P. *J. Mag. Res.* **1978**, *30*, 351
10. Garroway, A. N *J. Mag. Res.* **1982**, *49*, 168
11. Barbara, T. M. *J. Mag. Res. A* **1994**, *109*, 265
12. Bax, A.; Griffey, R.H.; Hawkins, B.L. *J. Mag. Res.* **1983**, *55*, 301
13. Griesinger, C.; Otting, G.; Wüthrich, K.; Ernst, R. R. *J. Am. Chem. Soc.* **1988**, *110*, 7870
14. Ganapathy, S.; Rajamohanan, P. R.; Ramanujulu, P. M.; Mandhare, A. B.; Mashelkar, R. A. *Polymer*, **1994**, *35*, 888
15. Kumar, A.; Ernst, R. R.; and Wüthrich, K. (1980) *Biochem. Biophys. Res. Commun.* **1980**, *95*, 1
16. Yu, K.-L.; Deshpande, M. S.; Vyas, D. M. *Tetrahedron Lett.* **1994**, *35,* 8919
17. Hiroshige, M.; Hauske, J. R.; Zhou, P. *Tetrahedron Lett.* **1995**, *36*, 4567

18. Sarin, V. K.; Kent, B. H.; Tam, J. P.; Merrifield, R. B. *Analytical Biochem.* **1981**, *117*, 147
19. Long, J. R.; Sun, B. Q.; Bowen, A. ; Griffin, R. G. *J. Am. Chem. Soc.* **1994**, *116*, 11950

Chapter 24

A High-Throughput Functional Assay for G-Protein-Coupled Receptors

Method for Screening Combinatorial Libraries

Christopher J. Molineaux, William Goodwin, Ling Chai, Xiaohua Zhang, and David I. Israel

Pharmaceutical Peptides, Incorporated, 1 Hampshire Street, Cambridge, MA 02139

We have developed a high-throughput functional assay in order to screen for novel ligands of G-protein-coupled receptors. This assay can be configured to identify either agonists or antagonists, and is sensitive to ligands at concentrations 10-20x less than the K_D for agonists, and 2-6x greater than the K_D for antagonists. This cell-based screen does not require a purified preparation of receptor or plasma membranes. We have employed the assay to screen peptide combinatorial libraries as well as to characterize lead compounds with respect to agonist and antagonist potency. For the receptors characterized thus far, the assay is sensitive, robust and reproducible and preserves the rank order of potency of known ligands. The assay is known as the MASTRscreen the Multiplex Assay for Seven Transmembrane Receptor agonists and antagonists.

Receptors in the seven-transmembrane (7-TM) superfamily are coupled to their effectors by way of guanine nucleotide-binding regulatory proteins (G-proteins) of various types, through which they respond to a wide spectrum of stimuli ranging from light to small molecule neurotransmitters, peptides and proteins (1-3). The family of 7-TM receptors is made up of at least several hundred distinct members and includes receptors that respond to signals ranging from environmental stimuli, such as photons, odorant molecules and sweet tasting sugars, to molecules involved in intercellular communication, such as biogenic amines, lipids, peptides and glycoproteins, including neurotransmitters, neuromodulators and hormones (4). Because of the involvement of G-protein coupled receptors in the regulation of many critically important biological functions, disease conditions may be influenced or determined by the state of activation or blockade of a G-protein coupled receptor. Receptor mutations responsible for human diseases are being elucidated at an increasing rate and may be either loss-of-function mutations or activating mutations. Examples include mutation of the thyrotropin receptor gene, leading to hyperfunctioning thyroid adenomas (5), mutation of the rhodopsin gene, leading to retinitis pigmentosa (6,7), mutation of the luteinizing hormone receptor gene, leading

1054–7487/96/0273$15.00/0

to precocious puberty (8) and mutation of the adrenocorticotropin receptor gene, leading to familial glucocorticoid deficiency (9). For a review of the role of 7-TM receptors in disease, see Coughlin (10). 7-TM receptors thus represent a significant pool of targets for the treatment of a wide range of human disease. As the understanding of receptor biology increases, it is becoming possible to design receptor ligands with correspondingly greater selectivity.

Given the important role of 7-TM receptors in both normal cellular responses and aberrant disease processes, assays that allow for the identification of agonists or antagonists of 7-TM receptors are highly desirable. Sensitive, selective and reliable high-throughput screens are now a key component of the drug discovery process. In the course of pursuing our drug discovery programs, we have developed, refined and optimized a functional assay for ligands of 7-TM receptors. The assay can be used to screen large libraries of compounds generated by combinatorial biology and chemistry and can also be used to characterize individual compounds to determine the relative potency and specificity of lead structures.

Materials and Methods.

Development of cell lines.   A cDNA clone encoding the human luteinizing hormone-releasing hormone (LHRH) receptor (LHRH-R) was kindly provided by Dr. J.L. Jameson (Northwester University Medical Center) and was identical to the previously published sequence (11), was evaluated for efficient expression in a number of host/vector system. Modified vectors designed for efficient expression of the LHRH receptor were evaluated by binding of radioligands after transient expression in COS cells transfected via DEAE-dextran. LHRH-R constructs were expressed in stable cell lines transfected using cationic liposomes and selected with either G-418 or methotrexate. Expression was confirmed using radioligand binding to membrane preparations or to whole cells. The M3 muscarinic receptor was expressed using similar methods.

Preparation of membranes from transfected cells: A washed suspension of cells was prepared in isotonic Hanks Balanced Salt Solution containing 20 mM HEPES buffer, pH 7.4 as described (12). After placement of 1-25 mL washed cell suspension in a nitrogen bomb (Gage) at 4°C with continuous stirring by a magnetic stir bar, the pressure was adjusted to 4-500 psi and left stirring for 20 minutes. Pressure was released and lysed cells were collected into a 15 or 50 mL plastic centrifuge tube containing a 100x cocktail of protease inhibitors (0.5 mM phenylmethylsulfonyl-fluoride (PMSF), 10 µg/mL benzamidine, 1 µg/ml leupeptin, final concentrations). The homogenate is centrifuged at 1000 g (~2500rpm) for 10 min. The supernatant was centrifuged at 10,000 g for 20 min. Pellets were washed with 2 mL membrane buffer and centrifuged at 10,000 g for 20 min, resuspended in 2.0 mL buffer and frozen in liquid N_2 in 50 µl aliquots. Aliquots of approximately 1 mg protein/mL were frozen in liquid N_2, stored at -80 °C and thawed at the time of assay. Recombinant receptors expressed in cell lines were used at 500,000-1,000,000 cell-equivalents/50 µL in the radioligand binding assay.

LHRH receptor displacement binding assay. Iodination of D-Ala⁶-LHRH-EA was accomplished using a modification of the Choramine T method (13). Briefly, 10 µL 100 µM D-Ala⁶-LHRH-EA was added to 70 µL 150 mM sodium phosphate, pH 7.4 and 10 µL Na¹²⁵I (1 mCi) in a 1.5 mL polypropylene tube. 20 µL Chloramine T (5 mg/mL) was added to initiate the reaction. After 5 min, 100 µL Na metabisulfite (1 mg/mL) was added and the entire reaction mixture was transferred to syringe barrel attached to a C18 SepPak module (Waters). Iodinated peptide was eluted by stepwise gradient of Methanol in 0.1 M acetic acid. Binding to the LHRH receptor in membrane preparations and whole cells utilized a modification of the assay according to Zhou et al. (14). 10 μL aliquots of test peptides (20x stock solutions) were put into individual wells of a 96-well polystyrene plate. 90 μL Binding Buffer (10 mM Tris, 0.1 % BSA, pH 7.4) was added to each well. 50 μL ^{125}I-D-Ala^6-LHRH-EA (30,000 cpm) in Binding Buffer was added to each well. 50 μL membranes or cell suspension was added to each well to start the binding reaction. The plate was incubated 90 min in ice and stopped by fast filtration binding using the Inotech 96-well cell harvester (Inotech). Filters were washed 3x with 300 μL 10 mM HEPES, 0.01% sodium azide. Filters were transferred to 12 X 75 mm tubes and counted for 1-5 min on the gamma counter (Packard). Non-specific binding was assessed by addition of 10 μM D-His^6-LHRH.

Binding assays with whole cells. To characterize receptor expression in recombinant cell lines, the binding of ligand to whole cells was used to determine approximate receptor number. This approach is rapid and sensitive, eliminates the need to prepare membranes and requires fewer cells than binding to membranes. Binding assays are performed with a protocol similar to that in membranes, with the exception that isotonic medium (Hanks/HEPES medium described above) was substituted for Binding Buffer.

MASTRscreen agonist functional assay. Cells expressing the appropriate G-protein linked receptor were seeded in 96-well plates as a subconfluent monolayer in 200 μL growth medium. An aliquot of 10 μL containing individual or multiple potential agonist compounds was added to the wells and serially diluted in half-log or log increments. After incubation for 3-5 days at 37°C in 5-10% CO_2, agonist activity was determined either qualitatively or quantitatively using dimethylthiazolyl diphenyltetrazolium bromide (MTT; Sigma, St. Louis, MO) using methods previously described (15,16). Briefly, 50 μL MTT (5 mg/mL in phosphate-buffered saline) was added and plates were incubated 2-4 hr at 37°C in 5-10% CO_2. The MTT substrate was aspirated and replaced with 200 μL isopropanol in 0.04% HCl. The blue formazan product was detected spectrophotometrically on a plate reader (BioRad) at 570 nm.

MASTRscreen antagonist functional assay. This assay is run similar to the agonist functional assay described above, except that potential antagonist compounds are added in the presence and absence of a stable agonist used at a concentration which gives a strong response in the assay.

Schild analysis and derivation of pA$_2$. Dose-response curves are generated by incubation of cells in the presence and absence of agonist and antagonist as described above. These data are replotted using Schild analysis to derive pA$_2$ values as a measure of antagonist potency (17). Agonist responses measured in the presence and absence of antagonist are plotted as a function of log agonist concentration. Curves are fitted to the logistic equation using a four-parameter fit according to the equation:

$$y = \frac{m1 - m4}{\left(1 + \left(\frac{x}{m3}\right)^{m2}\right)} + m4$$

where m1 = maximum absorbance
m2 = slope
m3 = EC$_{50}$
m4 = minimum absorbance

The rightward shift in the agonist response, referred to as the 'dose-ratio', is derived for each of the antagonist concentrations. These curves were used to derive the Schild plot according to the equation:

$$\frac{[A]_B}{[A]} = \text{log-dose ratio} = \log \left(1 + \frac{[B]}{K_B} \right)$$

where [A], $[A]_B$ = agonist concentrations inducing equal response in the
absence and presence of antagonist B, respectively
[B] = antagonist concentration
K_B = potency of antagonist B.

For each antagonist, the pA_2 value is the point where the corresponding curve crosses y=0, and can be derived from the equation:

$$pA_2 = -\log K_B$$

For antagonists that are strictly competitive with the agonist, the pA_2 value should roughly approximate the K_D determined by standard radioligand binding. Table 3 shows this to be the case for three of the four antagonists tested. One of these antagonists, DPhe^{2,6},Pro^3-LHRH, appears to be sensitive to some protease activity present in the MASTRscreen assay.

Results and Discussion.

The MASTRscreen assay has been used in both agonist and antagonist formats to identify and characterize ligands of 7-TM receptors. In both cases, a population of mammalian cells is incubated in the presence of test compounds for a period of 3-5 days. In the agonist format, the response to test compounds or combinatorial libraries of test compounds upon cells expressing the receptor of interest and the growth pattern is compared to that of cells which do not express the receptor. When the antagonist format is desired, incubation with the test compound is carried out in the presence of a specific, stable agonist of the receptor.

Agonist format. In the agonist format, the assay has been validated with a series of commercially available agonists for the LHRH-R, as shown in Figure 1. EC_{50} values for each agonist are shown in Table I. These values were approximately 10- to 20-fold lower than the corresponding K_D values (Table I), suggesting that induction of a response required about 5-10% receptor occupancy.

Antagonist format. In the antagonist format, the assay has been validated using several antagonists by their ability to block the response to the superagonist D-His^6-LHRH. Increasing concentrations of the antagonist Antide produced a progressive rightward shift in the dose-response curve to D-His^6-LHRH (Figure 2). The relative antagonist potency (pA_2 = 9.0), calculated by Schild analysis described in Materials and Methods (Figure 3), agrees well with the corresponding binding potency (pK_D = 9.5). Antide was completely inactive in the agonist format at all concentrations (data not shown). Three other commercially available antagonists have also been evaluated (Table II). Together, the results indicate that the assay is a sensitive and reliable method for evaluating the *in vitro* potency of antagonists active on this peptide receptor.

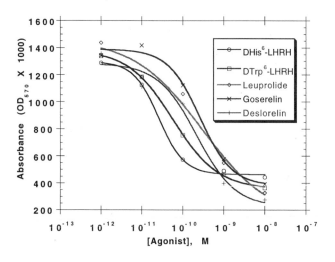

FIGURE 1. LHRH agonists produce a dose-dependent decrease in MTT activity. Increasing concentrations of LHRH superagonists were incubated in the presence of an LHRH receptor-expressing MASTRscreen cell line.

TABLE I. Binding *vs.* MASTRscreen response for the agonists shown in Figure 1.

PEPTIDE	BINDING	MASTRSCREEN RESPONSE	RECEPTOR OCCUPANCY
	IC_{50} (nM)	EC_{50} (nM)	
D-His^6-LHRH	0.2	0.03	12%
D-Trp^6-LHRH	2	0.06	3%
Leuprolide	6	0.3	6%
Goserelin	1	0.3	22%
Deslorelin	0.8	0.2	28%

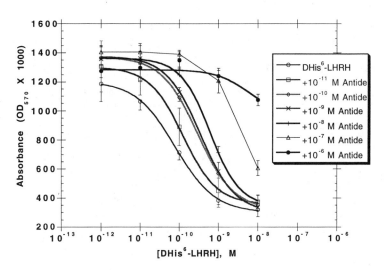

Figure 2. Antide blocks the agonist-induced reduction in cell proliferation. Antide, an antagonist for the LHRH receptor was evaluated at increasing concentrations. Progressive rightward shifts in the response curves for the agonist, D-His[6]-LHRH can be used to calculate the pA_2 value shown in Table II [DHis[6]-LHRH alone, N=8; all others, N=4].

TABLE II. Binding *vs.* MASTRscreen response for LHRH receptor antagonists.

PEPTIDE	BINDING	BINDING	MASTRCREEN
	K_D (nM)	pK_D*	pA_2
Antide	0.3	9.5	9
NacdhPro[1],pFPhe[2],DTrp[3,6]-LHRH	0.7	9	9
DPhe[2,6],Pro[3]-LHRH	7	8	<<7
DGlu[1],DPhe[2],DTrp[3,6]-LHRH	9	8	7

* pK_D = -logK_D

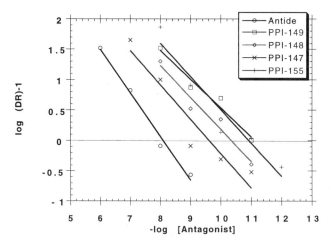

FIGURE 3. Schild plot of MASTRscreen antagonist responses.
Schild analysis of data from the assay, as described in the text.

The M3 muscarinic receptor is subjected to the MASTRscreen assay as shown in Figure 4. The muscarinic agonist carbachol induces a reduction in cell proliferation with an EC_{50} of approximately 0.1 μM. Although pirenzepine alone did not affect cell proliferation, addition of pirenzepine in the presence of carbachol produced a rightward shift in the agonist-induced dose-response curve.

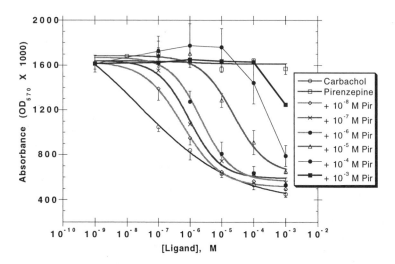

Figure 4. MASTRscreen antagonist assay for the muscarinic M3 receptor. The MASTRscreen signal induced by the muscarinic agonist carbachol is suppressed by the antagonist pirenzepine in a dose-dependent manner.

Simplified antagonist format

A simplified form of the MASTRscreen assay in antagonist format is shown in Figure 5. Here a range of concentrations of three proprietary antagonists has been assayed in the presence of a single concentration of D-His^6-LHRH, rather than a full matrix. While Schild analysis is not possible, this simplified form allows the potency of several different antagonists to be compared within a single assay.

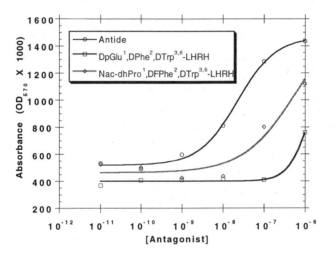

FIGURE 5. MASTRscreen assay in simplified antagonist format.
Each antagonist was evaluated in the presence of 1 nM D-His^6-LHRH rather than a full matrix.

For any given receptor, requirements of the assay are minimal and straightforward. The appropriate cDNA clone is necessary to prepare the recombinant cell line expressing the receptor, and for antagonist screening, a known agonist is required. Typically, the assay is carried out in a 96-well microtiter plate. For each test sample, the agonist format comprises two wells, one for agonist assay of the recombinant line and one for control assay of the non-recombinant parental line. The antagonist format comprises three wells, one for agonist assay, one for control and in addition, a third well that also contains the known agonist, for antagonist assay.

Screening of compound libraries to identify agonists and antagonists

The assay can also be used in 'screening mode' as a rapid, reliable, sensitive method for the detection of either agonists or antagonists. In a typical configuration, three 96-well microtiter plates are prepared, one each for antagonist format, agonist format and control. Each of the wells contains cells of the proprietary line expressing the appropriate target G-protein linked 7-TM receptor, and the compound pools to be screened. In the antagonist format plate, each well also contains a stable, well-characterized agonist. In the agonist format plate, the compound pools are screened directly without the addition of known ligands. Cells not expressing the recombinant receptor serve as specificity and toxicity controls. The plates are then incubated according to a defined cell growth protocol. After three to five days, a specific indicator is added that allows either simple visual readout or quantitative spectrophotometric measurement. Pools scoring positive can then be deconvoluted by iteration of the same procedure.

As a demonstration, we have evaluated a small combinatorial library of proprietary PPI peptides with the peptide receptor described above. Ten samples were assembled, each containing at least ten different peptides at concentrations of 1.5 μM each. In general the peptides in this library were chosen *not* to have activity on this receptor, except that one of the samples was spiked with the LHRH antagonist Antide and another with the LHRH superagonist D-His^6-LHRH.

Each sample was tested both in agonist format, as well as in antagonist format in the presence of 1 nM D-His^6-LHRH. Assays were carried out blind with respect to identity of the spikes. The results demonstrated that the agonist spike was clearly identified in sample 4, and the antagonist spike in sample 6 (Figure 6).

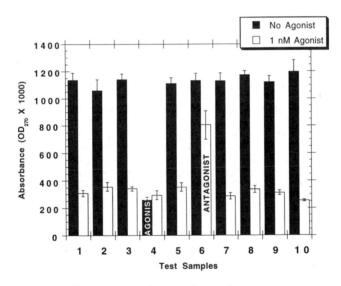

FIGURE 6. MASTRscreen assay in screening mode.
Ten samples of a small proprietary combinatorial library were spiked and tested blind in both agonist and antagonist formats, the latter in the presence of 1 nM D-His^6-LHRH. The assays clearly identify the spiked wells as agonist in sample 4 and antagonist in sample 6.

Receptor binding assays and *in vitro* functional assays are important tools in drug discovery. MASTRscreen offers several advantages over other assay systems. While receptor binding assays are quantitative, they do not distinguish between agonists and antagonists. Many functional screening assays now in use are unsuitable for primary high-throughput screening. The difficulties with current screening methods include the fact that they are too labor-intensive, the readouts are not suitable for high-throughput data acquisition and analysis, the expression systems are often unstable, and the coupling to reporter genes involve excessive gene manipulation. While a transformed cell assay *in vitro* cannot replace the eventual *in vivo* evaluation of a compound, the MASTRscreen assay is a sensitive, quantitative and reproducible non-radioactive method which has been shown to minimize or avoid the incidence of "false positives". It is not labor intensive, and is readily automated using currently available robotic procedures combined with a simple visual readout or spectrophotometric end point. The assay utilizes a stable proprietary expression system that does not require a reporter gene construct.

Acknowledgments

The authors would like to express their appreciation to Mr. Phillip Hamilton for help in preparing this manuscript and Dr. J. Larry Jameson at Northwestern University Medical Center for the LHRH receptor cDNA clone.

Literature Cited

1. Gilman, A.G.; *Ann. Rev. Biochem.* **1987**, *56*, 615-649.
2. Stryer, L., Bourne, H.R. *Ann. Rev. Cell Biol.* **1986**, *2*, 391-419.
3. Birnbaumer, L. *Ann. Rev. Pharmacol. Toxicol.* **1990**, *30*, 675-705.
4. Nathans, J. *Ann. Rev. Nuerosci.* **1987**, *10*, 163-194.
5. Parma, J., Duprez, L., Van Sande, J., Cochaux, P., Gervy, C., Mockel, J., Dumont, J., Vassart, G. *Nature* **1993**, *365*, 649-651.
6. Keen, T.J., Inglehearn, C.F., Lester, D.H., Bashir, R., Jay, M., Bird, A.C., Jay, B., Bhattacharya, S.S. *Genomics* **1991**, *11*, 199-205.
7. Robinson, P.R., Cohen, G.B., Zhukovsky, E.A., Oprian, D.D. *Neuron* **1992**, *9*, 719-725.
8. Shenker, A., Laue, L., Kosugi, S., Merendino, J.J., Minegishi, T., Cutler, G.B. *Nature* **1993**, *365*, 652-654.
9. Clark, A.J., McLoughlin, L., Grossman, A. *Lancet* **1993**, *341*, 461-462.
10. Coughlin, S.R. *Curr. Opinion Cell Biol.* **1994**, *6*, 191-197.
11. Flanagan, C.A., Becker, I.L., Davidson, J.S., Wakefield, I.K., Zhou, W., Sealfon, S.C., Millar, R.P. *J. Biol. Chem.* **1994**, *269*, 22636-22641.
12. DeMartino, J.A., Van Riper, G., Siciliano, S.J., Molineaux, C.J., Konteatis, Z.D., Rosen, H. and Springer, M.S. *J. Biol. Chem.*. 1994, *269*, 14446-14450.
13. Lasdun, A., Molineaux, C.J., Orlowski, M. *J. Pharmacol. Exp. Therap.* **1989**, *251*, 439-447.
14. Zhou, W., Rodic, V., Kitanovic, S., Flanagan, C.A., Chi, L., Weinstein, H., Maayani, S., Millar, R.P., Sealfon, S.C. *J. Biol. Chem.* **1995**, *270*, 18853-18857.
15. Mosmann, T. *J. Immunol. Methods;* **1983,** *65,* 55-63.
16. Hansen, M.B., Nielsen, S.E., Berg, K. *J. Immunol. Methods*; **1989**, *119*, 203-210.
17. Schild, H.O. *Br. J. Pharmacol.* **1947**, *2*, 189-195.

APPLICATIONS

Chapter 25

Use of Combinatorial Libraries in the Discovery and Development of Novel Anti-infectives

Mark A. Wuonola and David G. Powers

Department of Chemistry, Scriptgen Pharmaceuticals, Inc., 200 Boston Avenue, Medford, MA 02155

The aminoglycoside antibiotic neomycin B has been demonstrated to inhibit the association of the HIV encoded Rev-protein with a specific sequence of viral RNA termed the RRE (Rev response element) (*1*). Such inhibition could be therapeutically useful as both the RRE and the Rev protein are highly conserved, diminishing the potential of resistance through mutation (*2*). Traditional medicinal chemistry efforts were hampered due to the complex nature of the lead molecule, neomycin B. Simplified, more readily synthesized structures were targeted. We thus prepared directed combinatorial libraries based on 2-deoxystreptamine (2-DOS), the central residue common to most of the aminoglycoside antibiotics. Over 1000 mimics of neomycin B were generated in a short time; some posessed greater potency than the lead. We now consider 2-DOS to be a very promising scaffold for combinatorial chemistry directed toward not only antivirals, the area from which this work arose, but also for building universal libraries.

Advances in molecular biology, coupled with the advent of usable computer/robotics systems for screening, have necessitated the capability of more rapid analog preparation by the modern medicinal chemist. The answer, of course, has been the development of the field now termed combinatorial chemistry (*3,4*). The term "directed libraries" has been coined as a descriptor for compounds that are synthesized in parallel for a particular biological target based on the knowledge of a lead compound's structure, or a pharmacophore generated after the screening of a corporate library, whether random, small compound or combinatorial, for a given activity. It has become accepted that directed libraries will logically evolve as hits are found from screening of universal libraries. Many examples of this practice will doubtless be forthcoming as important collaborations of combinatorial companies and major pharmaceutical companies bear their fruit. The impact of this paradigm, however,

1054–7487/96/0284$15.00/0

depends on the versatility of the chemistry and on the true diversity of the original universal library. If the nature of substituents in the "universal" library is limited, then only incremental improvements will be possible through directed libraries. The more diverse the univeral library and the more imaginative the directed iterations, the more likely it is that significant further improvements will be made through directed approaches.

However, during our directed combinatorial efforts, based on a lead not derived from combinatorial chemistry, we gained an appreciation of our lead first as a directed combinatorial scaffold and then that it could be used to generate compounds for unrelated programs (*i.e.* generation of universal libraries). New universal libraries can stem from directed combinatorial programs. We have made some small combinatorial libraries in our efforts in both antiviral (Rev-RRE) and antibacterial (bacterial transcription inhibitors) programs. In both cases, our work in directed libraries around certain "hit" structures led us to an appreciation of the versatility of the *particular* scaffold for use in building *universal* libraries.

Chemistry Challenges of the Aminoglycosides

The structure of neomycin B shown in Figure 1 is typical of the family of aminoglycoside antibiotics that are substituted at the four and five positions of the central 2-deoxystreptamine (2-DOS) core. Neomycin B contains six amino groups, all of which can be positively charged at physiological pH (*5*). These amino groups clearly contribute to its known RNA binding (rRNA, RRE, TAR) but also to its very poor oral bioavailability and cell penetration, and are thought as well to contribute to the characteristic oto- and nephrotoxicity of the class (*6-9*). The number of functional groups and chiral centers made chemistry cumbersome. Even major efforts in the pharmaceutical industry have produced relatively few aminoglycoside analogs.

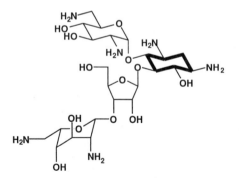

Figure 1. The structure of neomycin B. The 2-deoxystreptamine ring is shown in bold for clarity.

Total syntheses of neomycin B analogs were impractical for our medicinal chemistry program and would not afford a cost-effective drug (although total syntheses are reported in the literature) so preparation of analogs was limited to degradation or modification of the aminoglycoside and rebuilding, or semi-synthesis, to prepare desired analogs (10,11). The multiple similar functionality required elaborate protecting schemes. This, coupled with the difficulty in purifying and characterizing the complex, hydrophilic structures, severely limited the ability to obtain SAR knowledge by analoging or "traditional" medicinal chemistry. In addition, the poor oral bioavailability and cell penetration coupled with the specific toxicities, gave the series a number of hurdles to overcome which would prohibit rapid movement of a compound to the clinic.

2-Deoxystreptamine as a Combinatorial Scaffold

In the foregoing discussion, we gave the background of our discovery (on which a patent application has been filed) that 2-DOS, the central building block of the aminoglycosides, is an excellent combinatorial scaffold. Now will be described in detail how we conceived and synthesized a series of compounds with potential for inhibition of the Rev/RRE protein-RNA interaction. 2-DOS, whose structure is shown in Figure 2, is rigid, its conformation enforced by the equatorial disposition of the five substituents. These substituents can be independently functionalized, using chemistry which takes advantage of the symmetry of 2-DOS, and with five substituents, providing great combinatorial possibilities with even a rather small palette of functionalized pendant groups. Substituents can be arranged at certain of the five positions to probe varying regions of space and sites are still available for other substituents to add additional binding determinants (e.g., polar, uncharged or lipophilic) or to modulate the physical and ADME properties of the resultant molecule. Functionality is also available for attachment to a solid support when appropriate for combinatorial schemes.

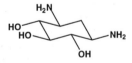

Figure 2. The structure of 2-deoxystreptamine (2-DOS).

The overall strategy that we envisioned for the combinatorial generation of neomycin B analogs based on 2-DOS can be seen in Figure 3. 2-DOS is prepared by degradation of neomycin B in better than 50% yield (12,13). It and related compounds have also been prepared synthetically but degradation is the most convenient way of obtaining it for laboratory studies. 2-DOS is readily converted to the dipyrrole, acetonide-protected form which is the starting point for functionalization with pendant groups. One pendant group can be introduced at the remaining hydroxyl, the acetonide protecting group is removed, additional groups are introduced on the unmasked hydroxyl groups, and the amidine functionality is introduced while the pyrrole protecting groups are removed. Other strategies utilize the dipyrrole-triol. Some of the arrays that were synthesized will be shown below.

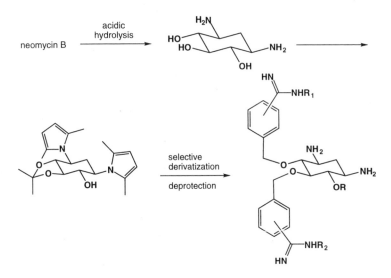

Figure 3. General stategy for the combinatorial generation of neomycin B analogs.

Combinatorial Synthesis of Amidine Mixtures

We chose amidines for our initial efforts because arginine, with its guanidine functionality, is so common in RNA-binding motifs of proteins. The DNA and RNA binding potential of *bis*-amidines has been recognized (*14,15*). As can be seen in Figure 4,

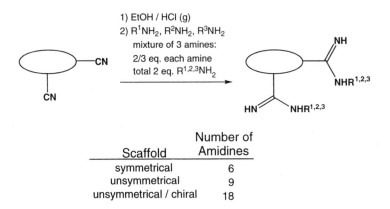

Scaffold	Number of Amidines
symmetrical	6
unsymmetrical	9
unsymmetrical / chiral	18

Figure 4. Use of the Pinner synthesis for combinatorial amidine generation.

by using the Pinner synthesis, the amidines can be made to bear aminoalkyl sidechains, modulating basicity and bearing additional positive charges. Depending on the structure of the starting nitrile, the use of a mixture of three amines in the reaction with the ethyl imidate provides 6, 9, or up to 18 different amidines. The reaction proceeds to completion with 1:1 stoichiometry, so even with a mixture of amines one can be sure that all the desired products will be present.

The initial libraries that were prepared were based on substitution of the 2-deoxystreptamine core with functionalized O-benzyl groups. Our choice of benzylic functionality stemmed from some preliminary modelling measurements that compared the distances between amine groups that had been shown to be important for binding in the lead neomycin, and the approximate distances between the terminal amidine functionality of our benzyl analogs. The procedure for the preparation of the bis-amidinobenzyl libraries, and several of the libraries to be described subsequently, began with the racemic acetonide protected dipyrrole 2-DOS synthon seen in Figure 5. Here, for example, the free hydroxy group is derivatized with picolyl or 3-nitrobenzyl chloride, the acetonide protecting group is removed, and the remaining two hydroxy groups are para-cyanobenzylated. All of the benzylations followed standard Williamson ether synthesis proceures and the products were purified by column chromatography.

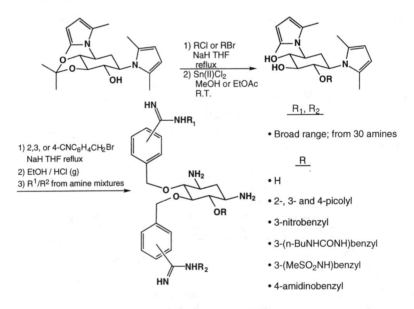

Figure 5. Synthesis of bis-amidinobenzyl analogs based on 2-DOS and the types of arrays prepared.

The benzyl function that was incorporated first, prior to acetonide removal and cyanobenzylation was included in these libraries to introduce added recognition elements into the structure in hopes of enhancing the association of the material to its' target relative to the lead molecule. The cyanobenzylated derivatives were then exposed to dry HCl gas in anhydrous ethanol to form an intermediate *bis*-ethylimidate salt which was isolated then added by syringe to mixtures of amines to complete the Pinner synthesis of the amidine functionality. The amidine hydrochloride salts precipitated from the reaction upon addition of diethyl ether, the solids were collected by centrifugation and decantation of the reaction solvent and by-products. The materials so obtained were then washed with the aid of a vortex mixer using diethyl ether and anhydrous ethanol sequentially. After each washing the solids were isolated by centrifugation and decantion. The final traces of sovent were then removed under reduced pressure. Using the procedure described above, over 1,000 variations were prepared in a short time using mixtures of three amines to prepare samples containing nine pairs of racemic amidines. Some compounds were also prepared as single compounds. All of these materials caused precipitation of RNA in the Rev/RRE filter-binding assay, showing that the materials bind strongly to RNA at submicromolar concentrations. Sequential removal of positively-charged groups should attenuate this effect allowing for continued affinity while enhancing specificity for the desired binding site. These arrays of compounds also show our ability to prepare materials containing both charged groups and also polar, uncharged pendant groups.

Having prepared materials that demonstrated very high, albeit non-selective, affinity for the RNA probe used in our biological assay, we wanted to ascertain whether adjustment of the distance between the charged amidine groups would elicit the desired characteristics. Consequently we prepared, using similar chemistry, analogs containing one *para*-amidinophenyl group and one *para*-amidinobenzyl group as shown in Figure 6.

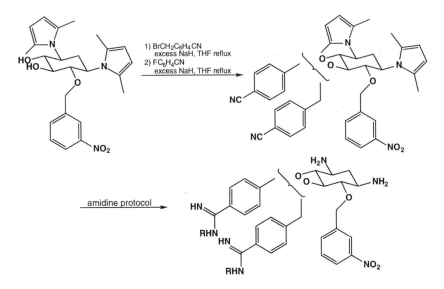

Figure 6. Synthesis of *bis*-amidinophenyl/benzyl analogs based on 2-DOS.

The key differences in this synthetic sequence in comparison to the previously described scheme are that a *para*-cyanophenyl fluoride was used rather than a benzyl halide, the second and third alkylations were performed sequentially, and column chromatography followed each of these sequential alkylations. As shown in Figure 7, we also prepared analogs where both of the amidinophenyl functions were appended through shortened linkages (*i.e.*, containing two *para*-amidinophenyl groups). These were prepared, using a slightly different synthetic sequence, from the dipyrrole triol by reaction with *p*-fluorobenzonitrile and chromatographic separation of the *bis*-cyanophenylated products, a mixture of 4,5- and 4,6-difunctionalized materials. These were converted into *bis*-amidines for evaluation by our standard amidine protocol.

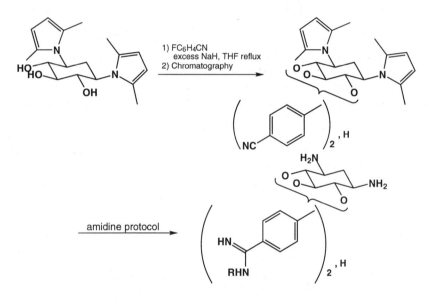

Figure 7. Synthesis of *bis*-amidinophenyl analogs based on 2-DOS.

Biological assay of the ability of these arrays to compete with Rev for its binding site on the RRE, in the presence of a large excess of *t*-RNAs as a test of specificity, showed that the compounds were now active and exhibited diminished propensity to precipitate the probe RNA. Also, the potency of the chain-shortened analogs was greater than that of the control neomycin B under the same conditions. The concentrations at which single compounds present in the mixtures inhibited the association of the protein by 50% ranged from 6 to 20 μM as compared to 30 μM for neomycin B. These concentrations were reaffirmed by bioassay of partially deconvoluted mixtures. The specific series that exhibited activity are shown in Figure 8.

Having prepared materials that were both simplified structurally and that posessed increased potency in a directed combinatorial approach afforded us a great deal of SAR information in a short period of time. This information has allowed us to design further simplified analogs, anticipated to have improved ADME properties. During the directed combinatorial program described above we became aware of the great combinatorial power and synthetic flexibility of the 2-deoxystreptamine scaffold.

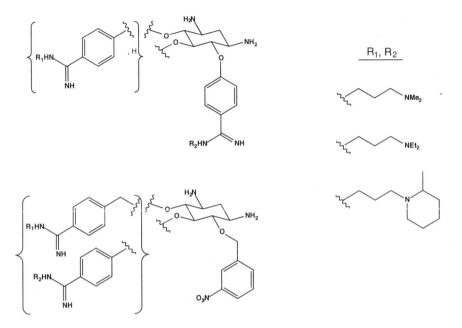

Figure 8. Chain length reduction evokes Rev/RRE activity.

Universal Libraries Based On 2-Deoxystreptamine

As a company we are interested in the development of small molecule, orally available drugs that can selectively inhibit clinically relevant protein-RNA interactions. To screen for compounds that possess these attributes, our corporate compound database requires molecules that display functionality conducive to RNA binding, and that are capable of recognizing the non-canonical structures adopted by RNAs that significantly differentiate them from other nucleic acid targets. Nature affords several instances of structural motifs used by proteins to accomplish this goal (*16,17*). The first and most commonly noted occurrence of such a domain is the arginine rich motif (ARM) found in RNA binding proteins. RNA binding sites of ARMs include stem-loops (N-proteins), internal loops (Rev) and bulges (Tat). The ARM motif obviously relies heavily on the association of positively charged residues on the protein side chain with the poly-anionic backbone of the RNA. It should be noted that this diverse selectivity is made possible by the presence of amino acids other than arginine in the ARMs. These include, in addition to the other basic amino acids lysine, and histidine, aromatic and polar-uncharged amino acids such as glutamine, asparagine, and tyrosine. Another RNA recognition motif termed the RGG box (ArgGlyGly) demonstrates that the modulation of the basicity of the side chains of the protein may allow for discrimination between nucleic acid targets. It is interesting to note that the RGG (ArgGlyGly) box often contains the post-translationally-modified amino acid, unsymmetrical dimethylarginine (DMA). This is thought to modulate specificity of RNA binding because the DMA is of increased size and hydrophobicity, but is still charged. Its hydrogen-bonding capability is altered, as well. It is also observed that the RGG units in RGG boxes are often interspersed with other, frequently aromatic (e.g. Phe, Tyr) amino acids. While charge-charge interactions may drive the associations in question, other factors have been demonstrated to be involved in the specific recognition of RNA structural motifs. The importance of shape complementarity, or preorganization of secondary structure, is demonstrated by the Rex nuclear localization signal. An interesting feature of the Rex nuclear localization signal is that in addition to nine basic amino acids in a 20 amino acid sequence, it contains six prolines. These both influence the conformation of this RNA-binding sequence and also are lipophilic. Also, it has been noticed that hydrophobic interactions play a role in the binding of RNAs by proteins, and may be the subtle force that is the key to specificity. The ribonucleoprotein (RNP) motif is comprised of 90-100 amino acids and occurs in proteins that bind pre-mRNA, mRNA, and pre-rRNA, and small nuclear (sn)RNA. The RNP motif contains two short highly conserved sequences. These sequences are often interspersed with other, conserved, mostly aromatic amino acids. These aromatic and lipophilic amino acids seem to be important as crystallographic information indicates that the aromatic amino acids are solvent exposed and make contact with RNA through ring-stacking interactions.

Libraries of compounds based on the 2-deoxystreptamine scaffold, incorporating facets of these known RNA binding motifs are appealing for our transcription targeted programs. Pendant groups that ought then be incorporated into the library based on this strategy include guanidine (as in arginine), aminopyridine, aminoalkyl, imidazole, benzimidazole, and pyridine functionality. These provide a range of basicity and represent groups that can be positively charged under physiological conditions. In accord with the other natural systems described above, polar-uncharged and lipophilic groups may strengthen and modulate RNA-binding, therefore additional pendant groups will be selected for inclusion in the palette to take advantage of binding potential of non-charged polar groups (e.g., benzamides, indole, benzothiazole, pyrimidines) and of lipophilic groups (alkyl, aryl, haloaryl, etc.). The structurally-characterized RNA-protein interactions contain some close contacts involving aliphatic amino acids, as well as of tyrosine. Additionally, phenylalanine, tyrosine, and tryptophan interact both as lipophilic groups and through pi-stacking. A few examples of RNA targeted pendant group functionality for 2-deoxystreptamine based libraries are shown in Figure 9 along with the amino acids which are models for each group.

Our initial efforts for the generation of RNA targeted universal libraries started with the amino acid functionalized derivatives as shown in Figure 10. These analogs can of course be prepared combinatorially using any of the known, commonly-used, peptide synthesis strategies. In these molecules the pendant groups can be derived from the natural amino acids as shown, the simplified ω–aminoalkanoic acids, or acyl groups bearing other basic substituents (pyridine, imidazole, amidines, etc.).

• <u>Polar, uncharged</u>   (Gln, Asn, Tyr, Trp)

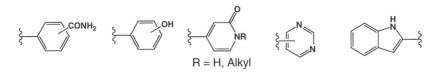

R = H, Alkyl

• <u>Aromatic</u>   (Phe, Tyr)

• <u>Lipophilic</u>   (Leu, Ile, Val, Phe, Pro)

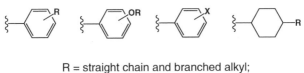

R = straight chain and branched alkyl;
X = halogen

Figure 9. Pendant functionality for RNA targeted libraries.

These molecules represent a step toward the more sophisticated RNA targeted structures suggested by the RNA-binding motifs discussed above. We call these prototypes because we are not restricted to using peptide-based pendant groups; we merely investigated these initially because of the availability of reagents used in peptide synthesis. It may be noted that hydroxyl groups have been used in other systems (steroidal scaffolds) for appending peptide groups and could be used analogously in our system to broaden the scope of the libraries prepared (*18*).

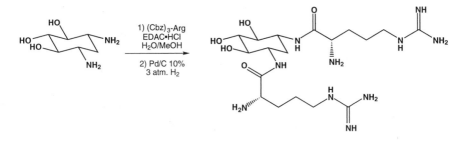

Figure 10. Prototype amino acid derivatives of 2-DOS.

The next phase of our investigation of available chemistries for the generation of 2-deoxystreptamine based libraries focused on acyl derivatives as shown in Figure

11. Some small arrays of *bis-N*-aroyl compounds have been made using aroyl and other acyl halides. Mixtures of three acyl halides were used to generate 6 compounds per well. QC by NMR and mass spectroscopy indicated that all the expected difunctionalized materials were present in representative wells. These should be viewed as prototypes for libraries in which nitrogens and at least one oxygen are functionalized, to take advantage of the combinatorial power of more than two variable pendant groups.

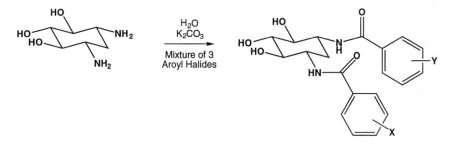

Figure 11. *N*-Aroyl functionalized libraries from 2-DOS.

Other chemistry that we have found to be useful for the preparation of analogs can be seen in Figure 12. Reductive amination of the amine groups of 2-deoxystreptamine can be performed in high-yield, making this dependable approach useful in a combinatorial scheme. An added feature of these analogs is that they are the reduced forms of the acyl derivatives described previously and contribute to the SAR picture by helping to determine the amount of preorganization required for specific recognition of the ligand for any target. Interesting extensions of this chemistry, for example when dialdehydes are used, include larger, dimeric or trimeric assemblies that may afford receptor libraries rather than ligand libraries (*18,19*).

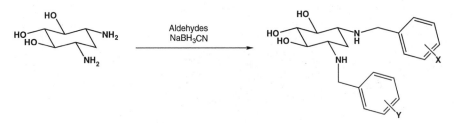

Figure 12. 2-DOS libraries prepared by reductive amination.

The next stage of our plan has been to incorporate what we had learned previously about the chemistry of the hydroxyl groups of 2-DOS from our *directed* combinatorial efforts with our ideas for amine functionalization for 2-DOS based *universal* library generation. For pragmatic reasons we have switched to solid support based chemistry as shown in Figure 13.

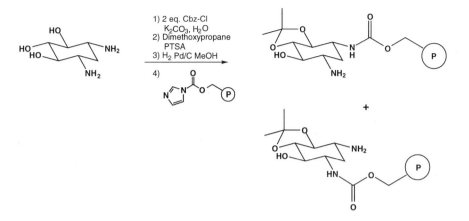

Figure 13. Attachment of a 2-deoxystreptamine intermediate to a solid support.

Solid phase synthesis is attractive for several reasons; low yield solution phase reactions can be forced to completion, it is ammenable to many of the reaction conditions that are of interst to us, and allows for the differential functionalization of the two amino groups in a regiochemically controlled fashion. Attachment of the 4,5-acetonide-2-DOS synthon to a polystyrene resin through a carbamate likage will provide the two regioisomerically substituted products shown above in Figure 13. However, our previous work, and others, have shown that 2-deoxystreptamine carbamates can undergo an intramolecular cyclization under mildly basic conditions to form an oxazolidin-2-one) as can be seen in Figure 14 (*20*).

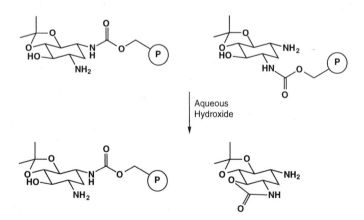

Figure 14. Strategy for regiochemical control of analogs prepared.

This reaction can only occur with the regioisomer that has a vicinal hydroxyl group allowing us to liberate one regioisomer and leave the other attached to the solid support. The material that remains on the resin is now regiochemically pure as a mixture of enantiomers. We are also studying enantiomeric control but cannot report any of our results at this point. The sequence of reactions to follow and some generic possibilities of N,N,O-trifunctionalized 2-DOS based libraries are illustrated in Figure 15. Functionalization at the unprotected amino group is followed by derivatization of the unprotected hydroxy group. Release from the solid support and functionalization of the second amino group ensues. The materials as shown are rather simple embodiments but serve to demonstrate the strategy. This combinatorial approach produces compounds that have pendant groups arranged about a central scaffold, however, the solid-support-based method can also permit us to extend the degree of oligomerization using *bis*-acyl donors (or aldehydes as mentioned above) and more than one 2-DOS component, to provide larger libraries that are likely to be much more suitable for specific binding to therapeutically-relevant RNAs.

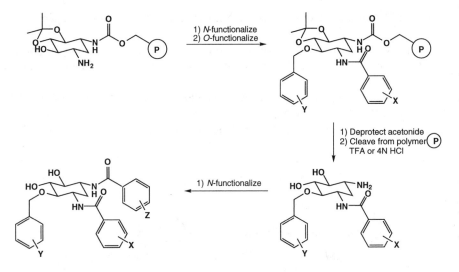

Figure 15. *N,N,O*-Trifunctionalized 2-DOS based libraries.

In summary, we have prepared a number of directed libraries based on the central structural unit of neomycin B, our lead compound in an HIV program targeting the Rev-RRE interaction. Some materials from these structurally simplified libraries were inhibitors of greater potency than the lead. In turn, the SAR information gained has assisted our traditional medicinal chemistry program in this area In the course of these directed efforts, the potential of the 2-deoxystreptamine scaffold for design of universal combinatorial libraries became evident. Our initial studies which form the basis for preparation of universal libraries by both solution and solid phase methodologies have been described. The libraries described have the ability to assume both the combinatorial approach of a functionalized central core and/or adopt an oligomeric theme. While our corporate targets are in the RNA area, the universal libraries contemplated, and the chemistry to prepare them, are not limited. Choices of pendant functionality can be made broadly for universal libraries and narrowed appropriately as directed libraries are contemplated for specific systems.

Acknowledgments

This work was supported by a phase I small business innovative research grant (SBIR) from the NIH - NIAID. The authors would like thank Professors James R. Williamson and Daniel S. Kemp for helpful suggestions. Methods development and library preparations were performed by Gary Gustafson, Zhe Li, Raksha Acharya, and Chi Tsung Choi. Biological assays and data are the work of James W. Lillie, Jaime Arenas, and Huiyun Zhang.

References

1.   Zapp, M.; Stern, S.; Green, M. R., *Cell*, **1993**, *74*, 969.
2.   Cohen, J., *Science*, **1993**, *260*, 1257.
3.   Ellman, J. A., *Chemtracts-Organic Chemistry*, **1995**, *8*, 1, and references contained therein.
4.   Desai, M. C.; Zuckermann, R. N.; Moos, W. H., *Drug Development Research*, **1994**, *33*, 174.
5.   Reid, D. G.; Gajjar, K., *J. Biol. Chem.*, **1987**, *262*, 7967.
6.   Foye, W. O.; Perlman, D., In *Principles of Medicinal Chemistry*, Lea & Feabinger Pub., 3rd Ed., Chpt. 33, pp. 695-697.
7.   *AMA Drug Evaluations*, 5th Ed., Sanders, W. B., Pub., pp 1242-1243.
8.   Prober, C., In *Principles of Medical Pharmacology*, B. C. Decker Pub.,**1989**, Chpt. 55, pp558-561.
9.   *Wilson and Gisvoldt's Textbook of Organic Medicinal and Pharmaceutical Chemistry*, 9th Ed., Delgado, J. M.; Remers, W. A., Eds., pp271-281.
10. Usui, T.; Umezawa, S., *J. Antibiotics*, **1987**, *40*, 1464.
11. Usui, T.; Umezawa, S., *Carbohydrate Research*, **1988**, *174*, 133.
12. Dutcher, J.; Donin, M.; *J. Am. Chem. Soc.*, **1952**, *74*, 3420.
13. Georgiadis, M. P.; Constantinou-Kokotou, V.; Kokotos, G.; *J. Carbohydrate Chem.*, **1991**, *10(5)*, 739-748.
14. Wilson, W. D.; Ratmeyer, L.; Zhao, M.; Strekowski, L.; Boykin, D., *Biochemistry*, **1993**, *32*, 4098.
15. Jansen, K.; Norden, B.; Kubista, M., .; *J. Am. Chem. Soc.*, **1993**, *115*, 10527.
16. Burd, C. G.; Dreyfuss, G., *Science*, **1994**, *265*, 615.
17. Hofer, L.; Weichselbraun, I.; Quick, S., King Farrington, G.; Bohnlein, E.; Hauber, J., *J. Virol.*, **1991**, *65*, 3379.
18. Boyce, W.; Li, G.; Nestler, P.; Sunaga, T.; Still, W. C., *J. Am. Chem. Soc.*, **1994**, *116*, 7995.
19. Goodman, M. S.; Jubian, V.; Linton, B; Hamilton, A. D., *J. Am. Chem. Soc.*, **1995**, *117*, 11,610.
20. Umezawa, S.; Takagi, Y.; Tsuchiya, T, *Bull. Chem. Soc. Japan*, **1971**, *44*, 1411.

Chapter 26

Application of Polymer-Supported Chemistry to the Discovery and Optimization of Lead Drug Candidates

A. L. Harris and B. E. Toyonaga

Ontogen Corporation, 2325 Camino Vida Roble, Carlsbad, CA 92009

Automated synthesis of small organic molecules using solid phase synthetic chemistry is a core technology at Ontogen. These libraries are prepared in 96 well reaction vessels (the OntoBLOCK system) designed to hold one reaction per well. Single compounds, spatially dispersed in the wells of 96-well plates are synthesized in high yield and with a high degree of purity. Bioactivity data, including IC_{50} determinations and preliminary kinetic parameters are rapidly obtained. Data analysis and compound tracking systems specifically designed for combinatorial chemistry and HTS have been developed and optimized for rapid data turnaround. Using these drug discovery tools, novel small molecules, exhibiting bioactivity against a number of molecular targets, have been identified in a short period of time. These include potent and selective, inhibitors of iNOS, PTPases and cdc25 phosphatase, as well as novel compounds that reverse the P-glycoprotein (Pgp) based multiple drug resistance (MDR) phenomenon in cellular assays and in animal models.

In the past, lead compounds have been discovered in the pharmaceutical industry through rational drug design or high throughput screening of either sample collections or natural product extracts. As a result of significant advances related to automated high throughput screening methods, the capacity to screen in biological assays far exceeds the output of traditional chemical synthetic methods employed by most medicinal chemistry departments or those that are available from natural product resources. In order to fill the need for larger, more diverse screening libraries, various methodologies are being developed to generate large numbers of small molecules using solid phase chemical syntheses techniques. Prior to the production of these libraries, a decision regarding compound synthesis strategy must be made. For example, compounds may be synthesized and tested as mixtures. Deconvolution of the active components is pursued after sample activity is identified. Deconvolution is achieved in many different ways, including the encoded bead methodologies (*1-2*). In contrast, other companies, including Ontogen, have chosen the spatially dispersed combinatorial library (SDCL) approach, producing single compounds after syntheses, rather than mixtures (*3*) (Mjalli, A. M. M.; Toyonaga, B. E. In *High Throughput Screening: The Discovery of Bioactive Substances* Devlin, J. P.; Wallace, R.; Marcel Dekker, Inc., New York, 1996, in press.).

The scope of this manuscript will be to describe the approach that Ontogen has taken to create SDCLs and the steps taken to discover biological activities and lead compound optimization.

Chemistry and Automation

Ontogen is a chemistry based drug discovery company accelerating the process of lead discovery and lead optimization by using proprietary automated methods of small molecule solid support chemical syntheses. Libraries of diverse compounds are prepared and tested in high throughput biological assays. The high throughput screens may be used for both lead discovery and lead optimization aspects of the drug discovery process.

A number of criteria must be satisfied in the design of a library prior to the initiation of methods optimization. A minimum of two to three chemical reagent inputs should be employed to produce a library of at least 50,000 compounds. In addition, it is desirable that all starting materials be commercially available, products be pharmaceutically attractive, methods be compatible with automation and approaches be novel and patentable. Finally, each compound must be synthesized in sufficient quantity for extensive testing, chemical analysis and archival storage. The synthesis flow chart for compound libraries using the OntoBLOCK system is summarized in Figure 1.

A schematic of the process used to meet the above criteria is summarized in Figure 2. Once compounds are synthesized using the OntoBLOCK system, they are re-distributed for storage (75-80%), high throughput screening (20-25%) and mass spectral (MS) analysis (~1%). The OntoMSPEC system uses electrospray mass spectrometry technology to test compounds directly from a 96-well microtiter plate format to achieve a throughput of up to 4 samples/minute. The user interface software is written in Visual Basic for MS Windows. It includes automated confirmation of expected ion(s). Display and query of MS data is integrated into the OntoQUERY software system for easy access to both biological and chemical information.

The OntoBLOCK system allows for the automation of a wide variety of chemistries, including multi-component condensation array (MCCA) synthesis, post MCCA transformations and general multistep linear synthesis (Mjalli, A. M. M.; Toyonaga, B. E. In *Exploiting Molecular Diversity and Solid Phase Synthesis*;

Library Synthesis

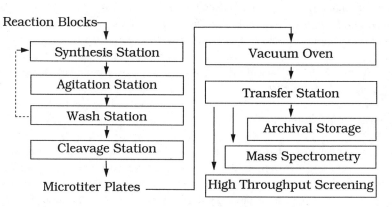

Figure 1. A schematic representation of the synthesis modules used for the creation of compound libraries using the OntoBLOCK system

Compound Library Planning, Synthesis and Analysis

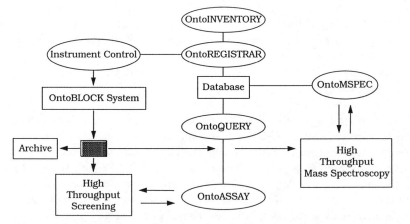

Figure 2. Some of the tools and the processes that are used at Ontogen to synthesize, analyze and screen chemical libraries. Software systems are outlined by ovals, while rectangles are used to denote hardware.

Chaiken, I.; Janda, K. Eds.; ACS Books, in press.). Several libraries using such diverse approaches have been synthesized at Ontogen (Table I):

Table I

- beta-lactams
- hydantoin imides and thioimides
- imidazoles
- indoles
- N-acyl-α-amino amides, esters, acids
- oxazoles
- phosphonates (α-hydroxy, α-amino, α-acylamino)
- phosphinates
- pyrroles
- tetra-substituted 5 membered ring lactams
- tetra-substituted 6 membered ring lactams
- tetrazoles
- thiazoles
- triazoles

These libraries have been evaluated in a variety of biological assays. Compounds showing biological activity have been optimized to have desirable pharmacological profiles using purified enzyme preparations, whole cell assays and *in vivo* animal experiments. Some of the results of such biological testing will be summarized in the next section.

Biology

A significant biology effort was initiated in order to discover and optimize lead compounds from the small molecule chemical libraries synthesized at Ontogen. This included the automation and validation of a number of primary screening assays in the search for novel inhibitors of therapeutically relevant molecular targets. Some of these molecular targets are listed below in Table II. Our early projects included the discovery of inhibitors of tyrosine phosphatases (PTPases), inhibitors of Pgp and inhibitors of iNOS. An account of the PTPase project has been published elsewhere (*4*) (Mjalli, A. M. M.; Toyonaga, B. E. In *High Throughput Screening: The Discovery of Bioactive Substances* Devlin, J. P.; Wallace, R.; Marcel Dekker, Inc., New York, 1996, in press.). This manuscript will focus on the latter two molecular targets and introduce the early results of a screening effort designed to discover novel inhibitors of the cell cycle associated enzyme cdc25 phosphatase. A summary of the progress in these aforementioned projects is summarized in Table III.

**Table II. Primary High Throughput Screening Assays Developed at Ontogen.
Asterisks (*) denote assays under validation.**

- cellular Pgp MDR
- cellular non-Pgp MDR
- CD45 (tyrosine phosphatase)
- cdc25 ("dual-acting" phosphatase)
- HePTP (tyrosine phosphatase)
- PTP1B (tyrosine phosphatase)
- iNOS (inducible nitric oxide synthase)
- cytokine receptor cellular assay
- Cathepsin B cysteine protease
- Pgp ATPase*
- Neuropeptide Y*
- F-1,6-bisPase*

Table III. Status of Selected Ontogen Drug Discovery Projects

Therapeutic Target	Project Goal	Current Status
P-glycoprotein	chemotherapy adjuvant	• multiple active classes • >100X more potent than verapamil or CsA • *in vivo* animal efficacy
PTPase X	modulation of signal transduction	• $IC_{50} \approx 100$ nM • >100X selectivity • competitive • non-phospho-tyrosine based
iNOS	anti-inflammatory	• $IC_{50} \approx 500$ nM • isoform specificity • non-competitive with Arg • irreversible • diaphorase inactive
cdc25	cancer treatment	• $IC_{50} \approx 3$ μM • cellular activity

Inducible Nitric Oxide Synthase (iNOS). The highly reactive free radical nitric oxide (NO) has been implicated in playing a variety of physiological and pathophysiological roles. NO is formed in the conversion of L-arginine to L-citrulline by the enzyme nitric oxide synthase (NOS) (5). Three major isoforms have been identified. Two are calcium/calmodulin-dependent, agonist-triggered, "constitutive" forms found in the brain and the blood vessel endothelium (denoted bNOS and eNOS, respectively). The third form is a cytokine-inducible form expressed in macrophages and other cells and termed iNOS (*6*).

A high throughput screening assay was designed following the methodology of Stuehr (*7*). Briefly, iNOS was isolated from γ-interferon and LPS-stimulated mouse macrophage cells (RAW264.7 cells). RAW cells were incubated overnight with γ-interferon and LPS. A lysate fraction was prepared and run over an ADP Sepharose column for partial purification. The assay was validated and optimized. Starting in April, 1995, randomly selected compounds from our libraries were initially tested at 100 μM in the primary assay. Within three weeks of screening our collection, at a rate of approximately 4,000 compounds a week, an active compound was identified, with an IC_{50} value of 23 μM. Subsequent studies demonstrated that this compound was not an arginine derivative and did not inhibit iNOS by competition at the arginine site. The chemistry group developed solid support synthesis methodology over the next two to three months and in July chemical synthesis was initiated. Within six weeks a series of compounds with IC_{50} values of 1 μM were identified. Within another three weeks a compound with an IC_{50} value of 0.5 μM was discovered. This drug discovery activity is illustrated in Figure 3. Compounds in this series were subsequently shown to have selectivity for iNOS versus bNOS (from rat brain). When tested against a panel of the three human NOS isoforms, compounds were identified with selective inhibitory activity against iNOS and others against bNOS. Finally, compounds in this series were shown to be devoid of nonspecific antioxidant effects, as demonstrated by their ability to retain their inhibitory activity in the presence of excess DTT (10 μM) and for their lack of inhibitory effect on another antioxidant sensitive enzyme, diaphorase (Lipoyl dehydrogenase).

Multiple Drug Resistance Reversing Agents. Multiple Drug Resistance is recognized as the leading cause of chemotherapeutic failure in cancer (*8*). A large percentage of patients exhibiting the MDR phenomenon have been show to overexpress an ATP-dependent, Mr 17,000 transmembrane protein known as P-glycoprotein (*9*). This protein serves to pump the cytotoxic chemotherapeutic compounds out of the cancer cells during treatment.

A number of commonly used chemotherapeutic agents are affected by this pump including the vinca alkaloids, vinblastine and vincristine, doxarubicin, etoposide and taxol. One current hypothesis to circumvent the MDR phenomenon is to inhibit Pgp by non-cytotoxic compounds, thus restoring the cytotoxicity of classical chemotherapeutic agents. A wide variety of known compounds have been shown to inhibit this pump, *in vitro*, including the calcium channel blocker, verapamil and the immunosuppressant, cyclosporin A (CsA). Unfortunately, the drugs inhibit the pump at concentrations that are at or above those needed to exert their original pharmacological effects, thus rendering them inappropriate for treatment of MDR.

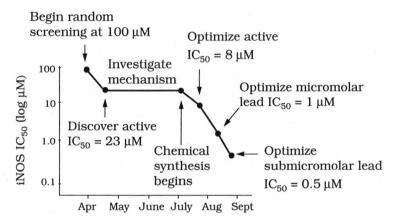
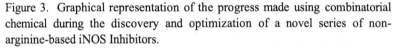

Figure 3. Graphical representation of the progress made using combinatorial chemical during the discovery and optimization of a novel series of non-arginine-based iNOS Inhibitors.

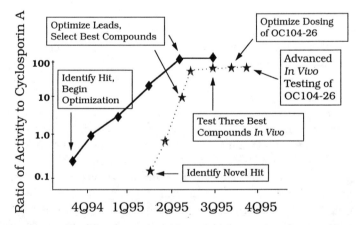

Figure 4. Graphical representation of the progress made using combinatorial chemical during the discovery, optimization and development of MDR reversing agents.

At Ontogen, a research program aimed at discovering novel, selective, orally reactive inhibitors of Pgp, and therefore MDR reversing agents, was initiated. A high throughput screening assay was developed to measure the accumulation of ^{3}H-vinblastine in human leukemia cells made resistant to vinblastine (CEM^{VLB1000}). Under basel conditions, very little ^{3}H-vinblastine is retained by these cells. In the presence of reference standard MDR reversing agents, verapamil and cyclosporin A, a concentration dependent 10-50 fold increase in the levels of ^{3}H-vinblastine occurs in these cells. Compounds that are active in this assay are then tested for their ability to enhance the cytotoxicity of vinblastine in the same cell line. Simultaneously, active compounds are tested to determine if they are intrinsically toxic on their own. Active compounds are then optimized for activity and lead compounds determined by testing them in a panel of MDR cell lines against a variety of cytotoxic agents. The best compounds from this discovery paradigm were then selected for *in vivo* efficacy studies. Figure 4 shows the progress we have made in this project by combining traditional pharmacological evaluation and combinatorial chemistry.

Two compounds, OC104-26 and OC42-92 have been shown to be 30-100 fold more potent than verapamil and CsA, *in vitro* (Figure 5) and have *in vivo* efficacy in a study using doxarubicin-resistant murine leukemia cells (P388^{ADR}) implanted in normal mice. OC104-26 was selected for further *in vivo* evaluation due to it's superior solubility when compared to OC42-92. The next study demonstrated that mice treated twice a day (50 mg/kg; BID) with OC104-26 extended survival to a greater degree than those mice treated with the same cumulative dose (100 mg/kg), given once a day (Figure 6). Finally, a pharmacokinetic study was performed to examine blood levels of OC104-26 after both i.p. and oral administration. As can be seen from Figure 7, concentrations of OC104-26 approached 5 µM at peak levels and maintained level at or above 1 µM for a minimum of four to eight hours after both routes of administration. These concentrations are well above the concentrations needed for *in vitro* efficacy. A summary of all the experimental results with OC104-26 can be found in below:

Summary of Pharmacology of OC104-26:

- *In vitro* cytotoxicity enhancement in resistant human cancer cell lines
- No intrinsic cytotoxicity in 11 different cell lines at 100 µM
- Minimal metabolism in two *in vitro* systems
- No intrinsic cytotoxicity in mice after 3 days at 100 mg/kg
- *In vivo* efficacy in a murine leukemia model (P388^{ADR})
- Improved efficacy with BID dosing
- Oral or i.p. administration result in significant plasma levels
- Inhibition of Pgp substrate transport (Hoechst 33342) at $IC_{50} = 0.4$ µM
- Inhibition of Pgp ATPase activity ($IC_{50} = 0.7$ µM)

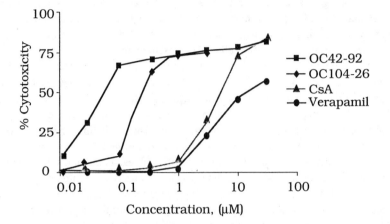

Figure 5. Cytotoxicity enhancement of vinblastine (VLB) in CEM[VLB1000] cells by OC90-42, OC104-26, cyclosporin A (CsA) and verapamil. Cells were cultured for 48 hours in the presence of 5 µg/ml vinblastine and various concentrations of reversing agent. Alamar blue was added and the cells were incubated for another 24 hours before reading on a fluorometer

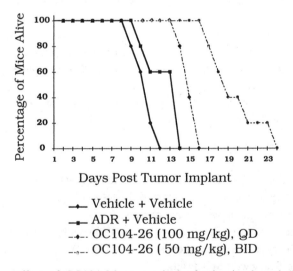

Figure 6. Effect of OC104-26 on survival of mice implanted (i.p.) with P388[ADR] (resistant) murine leukemia cells. Cells were implanted into the intraperitoneal cavity of the mice on day 0 and Adriamycin (2 mg/kg, i.p.) was administered on days 0, 4 and 8. OC104-26 (50 or 100 mg/kg, i.p.) was administered on days 0, 4 and 8 either once (QD) or twice a day (BID).

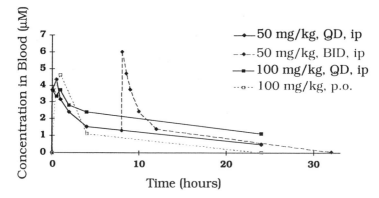

Figure 7. Concentration of OC104-26 in blood samples taken from normal mice at the denoted time points. Blood concentrations were determined on an HPLC. A minimum of three mice were used for each determination.

cdc25 Phosphatase. Cell cycle control mechanisms, while not fully elucidated, are becoming clearer through research on cyclins and cyclin-dependent kinases (cdks). It is now well established that the cyclin B/p34^{cdc2} and cyclin E/cdk2 kinases are necessary to bring the cell through the cell cycle and initiate proliferation. Before these complexes can become active kinases they must be dephosphorylated at adjacent sites containing tyrosine and threonine residues. A class of "dual acting" phosphatases, termed cdc25, are responsible for dephosphorylating both sites. It has been suggested that selective inhibitors of cdc25 phosphatase may be used as cytotoxic agents against cancer cells (*10*). We have successfully expressed cdc25A in an E. coli system and have characterized a cdc25A enzyme assay for high throughput screening. Randomly screening our library collection revealed a series of compounds with activity against cdc25 phosphatase with IC$_{50}$ values in the 2 to 30 μM range. These compounds are completely devoid of tyrosine phosphatase and serine/threonine phosphatase inhibitor activity. In subsequent studies, the ability of three of these compounds to inhibit growth of 14 different cancer cell lines was investigated. All three induced cytotoxicity of the cancer cells. This cytotoxicity was in the same rank order potency as their inhibitory activity against cdc25. Further studies are planned in order to determine if compounds are working via cdc25 inhibition to induce cancer cell death.

Conclusion

For the past two years, hardware advancements have been made in high speed chemical synthesis of spatially dispersed combinatorial libraries of small molecular weight compounds. Coupled to this has been the development and establishment of software systems to (i) assist in chemical library planning and registration, (ii) characterize chemical properties and (iii) analyze biological data. These approaches have allowed Ontogen scientists to rapidly identify novel, active, small (mol. wt. < 500) molecules inhibitors of a number of therapeutically relevant targets. Many of these molecules have undergone further optimization using our proprietary methods of highly efficient synthesis coupled with established medicinal chemistry strategies to generate interesting candidates for advanced preclinical development.

Literature Cited

1. Baldwin, J. J.; Bourbon, J. J.; Henderson, I.; Ohlmeyer, M. H. J. *J. Am. Chem. Soc.* **1995,** *117,* pp. 5588.

2. Janda, K. *Proc. Natl. Acad. Sci. USA* **1994,** *91,* pp. 10779-10785.

3. DeWitt, S. H.; Kiely, J. S.; Stankovic, C. J.; Schroeder, M. C.; Reynolds Cody D. M.; Pavia, M. R. *Proc. Natl. Acad. Sci. USA* **1993,** *90,* pp. 6909-6913.

4. Cao, X.; Siev, D.; Moran, E. J.; Lio, A.; Ohashi, C.; Mjalli, A. M. M. *Bioorganic & Med. Chem. Let.* **1995,** *5,* pp. 2953.

5. Feldman, P. L.; Girth, O. W.; Stuehr, D. J. *Chem. & Eng. News* **1993,** *51,* pp. 26-38.

6. Forstermann, U.; Schmidt, H. H. H. W.; Pollock, J. S.; et al. *Biochem. Pharm.* **1991,** *42,* pp. 1849-1857.

7. Stuehr, D. J.; Cho, H. J.; Kwon, N. S.; Weise, M. F.; Nathan, C.F. *Proc. Natl. Acad. Sci. USA,* **1991,** *88,* pp. 7773-7777.

8. Patel, N. H.; Rothenberg, M. L. *Invest New Drugs* **1994,** *12,* pp. 1-13.

9. Bradley, G.; Ling, V.; *Metastasis Rev.* **1994,** *13,* pp. 223-233.

10. Baratte, B.; Meijer, L.; Galaktinov, K.; Beach, D.; *Anticancer Research,* **1992,** *12,* pp. 873-880.

Chapter 27

Using Pharmacophore Diversity To Select Molecules To Test from Commercial Catalogues

Keith Davies

Chemical Design Ltd., Roundway House, Cromwell Park, Chipping Norton, Oxfordshire OX7 5SR, England

Compound library diversity is conveniently analyzed by considering the type and geometry of pharmacophores exhibited by the molecules in the library. This information may be applied to the selection of a representative subset of a combinatorial chemistry library or additional compounds from commercially-available catalogues for biological testing. Examples include the new *DIVERSet*™ and *HTS Chemicals* databases.

Pharmaceutical organizations currently test in the region of 20,000 molecules just to find one good lead for development into a marketable drug. These compounds come from different sources, whether synthesized by traditional methods and held in in-house archives, created using combinatorial chemistry methods, or else selected from the growing number of commercial catalogues now available. Indeed, the use of a variety of sources is recommended to ensure a wide diversity of molecules are considered. Effective design of compound libraries can reduce the number of compounds which need to be made or bought, without decreasing the diversity of the testing library. This has the potential of finding leads more rapidly because a smaller number of molecules need to be tested for specific activity, avoiding molecules which are very similar to each other and those previously tested.

This paper describes a new approach to analyzing library diversity by considering the type and geometry of pharmacophores which are exhibited by the molecules in a library, and how this information may be successfully applied to the selection of additional compounds, for example from commercial catalogues such as *DIVERSet*™ [1] and the databases contained on the *HTS Chemicals* CD-ROM [2].

The same approach can also be used to select representative subsets of combinatorial chemistry libraries.

The use of pharmacophores to describe the minimum requirements for activity is already proven for successful lead generation using 3D database searching. The same innovative pharmacophore technology has now been applied to diversity analysis within Chemical Design's established *Chem-X* software and is fully integrated within an extensive suite of lead generation, lead explosion and lead optimization tools for drug discovery.

Drug Discovery Architecture

The pharmacophore diversity tools present in *Chem-X* [3] form part of a comprehensive modular solution for combinatorial chemistry (see **Figure 1**), integrating library registration and design with links to robotics and biological test results.

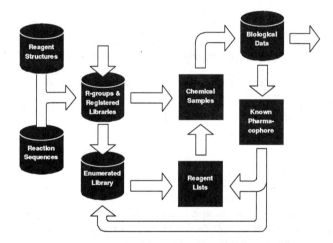

Figure 1. An information architecture for combinatorial chemistry

At each stage in the overall scheme, Chemical Design's integrated software tools make a significant contribution to drug discovery.

Library Registration. On registration a combinatorial library is recorded in terms of component parts or R-groups. These may be drawn in or generated from structure databases using reagent searching. *Chem-X* is also able to store generic reaction sequences from which the reagent queries and R-groups are generated automatically.

Library Design. Pharmacophore diversity analysis offers a systematic method for selecting molecules based on previous selections and activity information. This means a subset of a library can be chosen for testing which represents all pharmacophores (as described later in this paper). Chemical Design's pioneering pharmacophore diversity software is unique in providing this capability.

Lead Generation. For lead generation, it is desirable to test molecules with dissimilar frameworks for biological activity. Eliminating molecules with similar R-groups is therefore performed as an initial step to ensure that the molecules analyzed for pharmacophore diversity are dissimilar. The industry standard Tanimoto similarity index is used based on a fingerprint derived from the 2D database keys. In addition, *Chem-X* provides two methods of selecting dissimilar R-groups: diverse ordering and similarity clustering.

Pharmacophore diversity analysis then reduces the number of compounds which need to be made and tested by selecting the subset which covers the maximum diversity of pharmacophores in the minimum number of molecules.

Lead Explosion. The pharmacophores identified from active molecules are used to generate a focused library where each molecule shares pharmacophores with active molecules. This enables molecules with the desired pharmacophores but with alternative frameworks to be made and tested, thus producing many more leads.

Lead Optimization. For lead optimization, once an active pharmacophore is identified, derivatives of leads are investigated by systematically exploring R-group property space. For some properties such as molecular volume, there is normally a constant difference between the values for an entire series. When chemists are selecting reagents it might be preferable to calculate and store in the reagent database a set of properties which can be used by anyone to make selections. Users creating component databases from which libraries are registered will also need to calculate properties if R-group property selection is to be used.

Definition of a Pharmacophore

The simplest definition of a pharmacophore is the minimum features required in a molecule for activity. Most published examples consist of 3 or 4 essential interactions, such as hydrogens bonds, charge interactions, hydrophobic interactions and pi-interactions. In *Chem-X* a pharmacophore is described in terms of the interaction centers involved and the distances between them (see **Figure 2**).

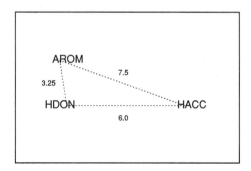

Figure 2. Example pharmacophore with one aromatic center and hydrogen bond donor and acceptor centers.

Following previous work [4], seven center types are now considered for pharmacophore identification:

- Hydrogen bond donors
- Hydrogen bond acceptors
- Positively charged centers
- Aromatic ring centers
- Hydrophobic centers
- Acidic centers
- Basic centers

Any molecule which exhibits the required pharmacophore would be expected to be active, although for high activity other groups may be required. For example, an extended ring system could increase activity from micromolar to nanomolar levels.

Quantifying Diversity

Pharmacophore diversity can be quantified in terms of the number of pharmacophores exhibited by a library. In this work 3-center pharmacophores are considered and there are 84 ways of selecting them from the 7 center types. In *Chem-X*, a distance bin model is used for distances in the range 0-15 Ångstroms of 31 bins. This means that there are 29791 logical triangles (although some of these are imaginary because, for example, the long side is longer than the sum of the two shorter sides). A pharmacophore key (Figure 3) is constructed from the list of pharmacophores where each bit corresponds to a specific selection of 3 centers, *e.g.* Donor, Acceptor, Aromatic ring and a precise geometry *e.g.* 6.0, 3.25, 7.5. These keys use approximately 312Kbytes each.

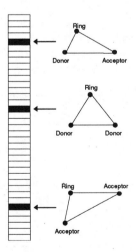

Figure 3. Pharmacophore key

Calculating and visualizing pharmacophore keys

In order to calculate the pharmacophore key for a library, the **Chem-X** combinatorial chemistry library builder generates 3D coordinates. Then an automatic conformational analysis for each molecule identifies rotatable bonds and uses rules to predict and skip conformers with a high energy [5]. The bits in the pharmacophore key are set for the pharmacophores exhibited by each low energy conformer.

A subset of the library is selected by considering each molecule in turn. If the Nth molecule exhibits pharmacophores which largely overlap with those exhibited by molecules previously included in the subset, then the molecule is not selected for inclusion in the subset. The degree of overlap may be specified by the user. Very rigid and very flexible molecules are also normally excluded. The pharmacophore key is not usually stored per molecule but is routinely stored for each library or mixture.

It can often be useful to visualize the pharmacophore key as a 3-dimensional plot (**Figure 4**) where the X,Y,Z axes are used for the distances between the centers. Symbols are used to indicate the various selections of centers and colors may also be used to code the symbols by some property.

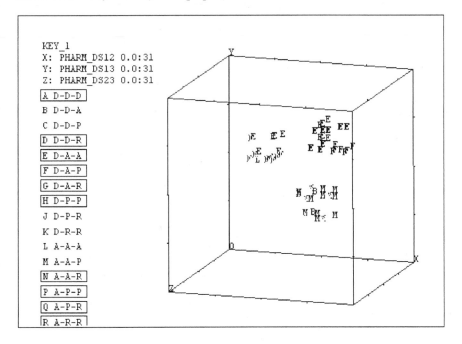

Figure 4: Pharmacophore plot for tyrosine

In this pharmacophore plot each symbol or letter corresponds to a discrete pharmacophore. The legend indicates which letters correspond to each type of pharmacophore. Using various tools, symbols may be selected and the corresponding pharmacophore displayed or saved for use as a search query or in another pharmacophore key.

Determining the Overlap of Libraries

The number and percentage of set bits in the pharmacophore key may be reported to users of the software giving a quantitative measure of diversity. By performing logical operations such as AND, NOT, OR and EOR the overlap and differences between pharmacophores exhibited by libraries may be readily determined. A tolerance is used to compensate for rounding errors for distances at or close to the bin boundaries and to accommodate the variation in distance that may be acceptable to the receptor. This approach may determine whether it is useful to make or purchase a library for testing. A library which exhibits pharmacophores largely similar to those exhibited by libraries previously tested is unlikely to generate new leads.

Example of Library Comparison

In this work two libraries were considered - a tri-peptide library using 26 amino acids in each position and a beta-lactam library using 26 amino acids, 20 aldehydes and 13 acid chlorides. Both libraries were registered and enumerated, and were compared on the basis of the pharmacophores exhibited.

A systematic rule-based conformational generation approach was used for molecules with up to 8 rotatable bonds and random-rule-based for more than 8 bonds. The rule-based approach eliminated conformers which are predicted to have high energy prior to generating the coordinates. A summary of the comparison is given in **Table 1.**

Table 1. Diversity analysis of two registered combinatorial libraries

	Number of compounds	% of available pharmacophores	Time/hr
Beta-lactam	6760	62.1	12
8-bond subset	304	25.7	7
12-bond subset	539	54.7	10
Tri-peptide	17576	83.7	41
8-bond addition	28		4
12-bond addition	535		12

The simplest comparison of the libraries is to compare the number of pharmacophores exhibited. In this case the number of pharmacophores is expressed as a percentage of the number of possible pharmacophores, 62.1% for the beta-lactam library and 83.7% for the tri-peptide library.

A more meaningful comparison can be made by considering the differences in geometry of the pharmacophores. Using the standard *Chem-X* tolerance (0.5 A), a logical NOT operation identifies 32821 pharmacophores which are exhibited by the tri-peptide library which are not exhibited by the beta-lactam library.

A diverse subset of the beta-lactam library was selected containing 539 molecules from the 6760 in the library. Constraints eliminating molecules which exhibited more than 100 000 pharmacophores or more than 12 rotatable bonds were used. The resulting set exhibited 293 027 pharmacophores of the 332 832 exhibited by the entire library. On a Silicon Graphics Indigo 2 this calculation took 10.3 hours.

The tri-peptide library was also analyzed to identify molecules which would add to the diversity of the beta-lactam library. This calculation was performed with a maximum of 8 and 12 rotatable bonds selecting 28 and 535 molecules respectively.

DIVERSet™ from ChemBridge

The benefit of using pharmacophore diversity to select a subset of compounds from a commercial collection for testing at the lead generation stage is exemplified by *DIVERSet*™. This library of hand-synthesized small molecules was rationally selected from the vast ChemBridge compound collection to cover the maximum pharmacophore diversity in the minimum number of molecules. The selection process featured Chemical Design's pharmacophore diversity analysis as described above and the result is a diverse subset of molecules highly suitable for use in screening programs (see Table 2 for a summary of *DIVERSet*™):

Table 2. *DIVERSet*™ summary (1995 figures)

Number of compounds	Number of pharmacophores	
10,000 (1995)	200,000 (1995)	>1 molecule per pharmacophore

HTS Chemicals on CD-ROM

Agreements with leading chemical suppliers have also enabled Chemical Design to collect their compound databases on a single CD-ROM (see **Table 3**) which can be searched by 2D/3D structure and data field. For each database the diversity of the molecules it contains is expressed using a pharmacophore key. Comparison of keys enables compounds with new or untested pharmacophores to be selected for purchase to complement existing databases.

Table 3. Content of *HTS Chemicals* (JAN96 release)

Supplier	Number of compounds
ChemBridge Corporation	52473
ChemStar	40833
ComGenex	39685
Contact Service Co.	34473
Maybridge Chemical Co.	49812

Conclusions

Pharmacophore diversity analysis provides an efficient means of comparing structurally diverse libraries, selecting subsets of the libraries for testing, and generating selections which complement previously tested molecules. This functionality is present as part of an integrated combinatorial chemistry software solution in *Chem-X*.

References

1.   *DIVERSet*™ is a diverse collection of compounds for high-throughput screening, selected from the ChemBridge compound collection. ChemBridge Corporation, 5 Revere Drive, Northbrook, Illinois, USA.
2.   *HTS Chemicals* is a collection of commercial compound databases for high-throughput screening. Chemical Design, Roundway House, Cromwell Park, Chipping Norton, Oxon, UK.
3.   Chem-X Reference Manuals, Chemical Design (as above).
4.   Martin, Y.C. et al., *J. Comput. Aided Mol. Des.*, **1988**, *2*, 15.
5.   Murrall, N W and Davies, E K., *J Chem Inf. Comput. Sci.* **1990**, *30.3* 315.

Author Index

Salemme, F. R., 16
Salvino, J. M., 16
Shen, Yeelana, 10
Shi, S., 219
Sia, Charles, 2
Soll, R. M., 16
Spurlino, J. C., 16
Stankova, Magda, 137
Strop, Peter, 137
Subasinghe, N., 16
Tarby, Christine M., 81

Tartar, André, 255
Thomas, Bert E., 10
Thompson, K. A., 158
Tomczuk, B. E., 16
Toyonaga, B. E., 70,298
Urban, Jan, 2
van Eikeren, Paul, 199
Weichsel, Aleksandra S., 99
Wright, Peter, 199
Wuonola, Mark A., 284
Zhang, Xiaohua, 273

Affiliation Index

Affymax Research Institute, 58
Alanex Corporation, 219
Argonaut Technologies, Inc., 199
Bohdan Automation, Inc., 188
Chemical Design Ltd., 309
ChromaXome Corporation, 158
CombiChem, Inc., 81
Connaught Laboratories, 2
3-Dimensional Pharmaceuticals, Inc., 16
Diversomer Technologies, Inc. 207
EnzyMed, Inc., 144
Institut Pasteur de Lille, 255
Isis Pharmaceuticals, 40
MDL Information Systems, Inc., 233
Molecumetics Ltd., 2
Ontogen Corporation, 70,298
Parke-Davis Pharmaceutical Research, 207

Pharmaceutical Peptides, Incorporated, 273
PharmaGenics, Inc., 30
Procept Inc., 10
Scripps Research Institute, 81,118
Scriptgen Pharmaceuticals, Inc., 284
Selectide Corporation, 99,137
SmithKline Beecham Research and
Development, 172
Texas A & M University, 128
Torrey Pines Institute for Molecular
Studies, 50
University of California—Berkeley, 144
University of Iowa, 144
University of Pennsylvania School
of Medicine, 172
University of Washington, 2
Zeneca Pharmaceuticals, 246

Subject Index

A

α helices, role in biological recognition,
172–173
N-Acylamino ethers, synthesis of
library, 113,114f
Adenosine
derivation, 147,150f
derivative structures, 147,152–153
structure, 147,148f
Affinity element(s), description,
222

Affinity element analysis, 225–226
Agonist format, high-throughput
functional assay for G-protein-coupled
receptors, 276,277f,t
N-(Alkoxyacyl)amino acids, synthesis
of library, 113,114f
N-(Alkoxyacyl)amino alcohols,
synthesis of library, 113,114f
N-(Alkoxyaryl)diamines, synthesis of
library, 113–115
N-Alkylamino ethers, synthesis of
library, 113,114f